EUSTON

EUSTON

A History and Modelling the 1875 Station

David Ashwood

&

The Market Deeping Model Railway Club, CIO

AN IMPRINT OF PEN & SWORD BOOKS LTD.
YORKSHIRE – PHILADELPHIA

First published in Great Britain in 2024 by
Pen and Sword Transport
An imprint of
Pen & Sword Books Ltd.
Yorkshire - Philadelphia

Copyright © David Ashwood, 2024

ISBN 978 1 39908 171 9

The right of David Ashwood to be identified as author of this work has been asserted by him in accordance with the Copyright, Designs and Patents Act 1988.

A CIP catalogue record for this book is available from the British Library.

All rights reserved. No part of this book may be reproduced or transmitted in any form or by any means, electronic or mechanical including photocopying, recording or by any information storage and retrieval system, without permission from the Publisher in writing.

Typeset in 11.5/14 Palatino
by SJmagic DESIGN SERVICES, India.

Printed and bound by Printworks Global Ltd, London/Hong Kong.

Pen & Sword Books Ltd. incorporates the imprints of Pen & Sword Books: After the Battle, Archaeology, Atlas, Aviation, Battleground, Discovery, Family History, History, Maritime, Military, Naval, Politics, Railways, Select, Transport, True Crime, Fiction, Frontline Books, Leo Cooper, Praetorian Press, Seaforth Publishing, Wharncliffe and White Owl.

For a complete list of Pen & Sword titles please contact

PEN & SWORD BOOKS LIMITED
George House, Units 12 & 13, Beevor Street, Off Pontefract Road,
Barnsley, South Yorkshire, S71 1HN, England
E-mail: enquiries@pen-and-sword.co.uk
Website: www.pen-and-sword.co.uk

or

PEN AND SWORD BOOKS
1950 Lawrence Rd, Havertown, PA 19083, USA
E-mail: uspen-and-sword@casematepublishers.com
Website: www.penandswordbooks.com

Contents

Acknowledgements 7

1 Introduction 8
2 Why Was Euston Built? 17
3 Railway Development and Euston 22
4 The Euston Grove Vista 28
5 The Station Buildings 46
6 The Platforms 58
7 The Railway Clearing House 73
8 Under Euston 82
9 The Locomotive Sheds 87
10 The End and a New Beginning 94
11 The Sultan of Zanzibar's Visit 105
Appendix A   The Unratified Anti-slavery Treaty 110
Appendix B   Slavery Treaty Ratification 112
Appendix C   *Hansard* Parliamentary Extracts for the Sultan's visit 113
Appendix D   *Hansard* Parliamentary Extracts for the Euston Area 115
Appendix E   *Hansard* Parliamentary Extracts for the Building of the London and Birmingham Railway 116
Appendix F   1946 Ordnance Survey Photo Mosaic Element for Euston East 117
Appendix G   1946 Ordnance Survey Photo Mosaic Element for Euston West 118
Bibliography and Useful Sources 119
Why we 'do' Model Railways 120

Acknowledgements

Our thanks to Pen & Sword Books for taking the brave move to involve another facet of the modelling hobby beyond the military and technical. Their extensive coverage is sure to tempt anyone who has looked twice, longingly, at a museum exhibit or model kit.

Thank you also to members of the Market Deeping Model Railway Club for their kind assistance and willingness to share and advise, particularly to Alan Hancock and Peter Davies for their help in reviewing and editing early drafts of this book.

Historical photographic assistance is essential for a book such as this. We would like to acknowledge the personal help in sourcing images from the Online Transport Archive and Friends of Lens of Sutton. The LNWR Society archives have likewise proven to be a invaluable source. The archaeological and other images of the HS2 site excavations have been sourced via that company's press office. Thank you to all the above for your assistance. Also, to acknowledge Ben Brooksbank for placing so many great images in the public domain.

1
Introduction

Typical of the post-war Euston expresses. BR (LMR) standard 4-6-0 Britannia No. 70044 *Earl Haig* heading the 'up' 'Mancunian' near Watford on 19 March 1955. As yet unnamed, since this locomotive (along with sister No. 70043) had just received smoke deflectors following a trial from new, with twin front-mounted Westinghouse air pumps. (*Online Transport Archive AND-515*)

'Railway termini are our gates to the glorious and the unknown. Through them we pass out into adventure and sunshine, to them, alas! we return.'

E.M. Forster

The Market Deeping Model Railway Club was formed in 1976, the year of a prolonged and notorious drought and heatwave. The Club is a thriving social ecosystem of like-minded railway modellers, with a desire to learn, share, specialise and display the end results to the public at exhibitions. The Club members are a wonderful mix of different careers, from professional railway people through to an ex-bank manager, computer support people and a carpenter. It has very much a 'can do' attitude and a common love of building and operating layouts.

Disaster struck the Club on the night of 17 May 2019. The Annual Model Railway Show had been set up at a school that evening in the ancient town of Stamford, Lincolnshire, all ready for an early start the next morning. Layouts were set up and tested, 'position one' rolling stock was in place. Traders had arrived and their wares were set out ready to sell.

During the night the premises was invaded, and an extensive spree of vandalism occurred. The next morning the author and his wife, who were expecting to make the morning's breakfast rolls, opened up to discover most of the contents had been comprehensively reduced to matchwood. Our own layouts, those of fellow clubs and the traders' stalls were all demolished.

Everyone was shocked. In some cases, twenty-five years of work was gone; mature tears abounded. The Club and all others present experienced shock, anger and disbelief. That Saturday afternoon was spent with brooms and dustbins, and we wondered just where we would go from here to offset the damage and loss of Club earnings. We decided to set up a £500 'Just Giving' request online to reduce losses.

During the following week, news of the vandalism was flashed around the world, seizing the common imagination. Club members appeared on television and radio. Our one operable locomotive, taken from a raffle prize, was set up on our Club's test track to show some background movement. Members of the railway modelling fraternity, wargamers, the general public, people with fond memories of their grandfathers' past, provided kind words, and donations of all kinds poured in. From Miniatur Wonderland of Hamburg and Sir Rod Stewart through to children's pocket money and a lady from Japan apologising for her English. It should be realised that not only the financial investment, but the time and devotion of past and older members, had been utterly lost.

As a result the Club decided to become a charity, to process those donations appropriately, curate a historical collection of assets representing the evolution of the hobby in Britain and elsewhere, promote model clubs for local children and assist other local good causes. Good can come from bad, eventually. Our share of profits from this book goes directly back into the charity fund.

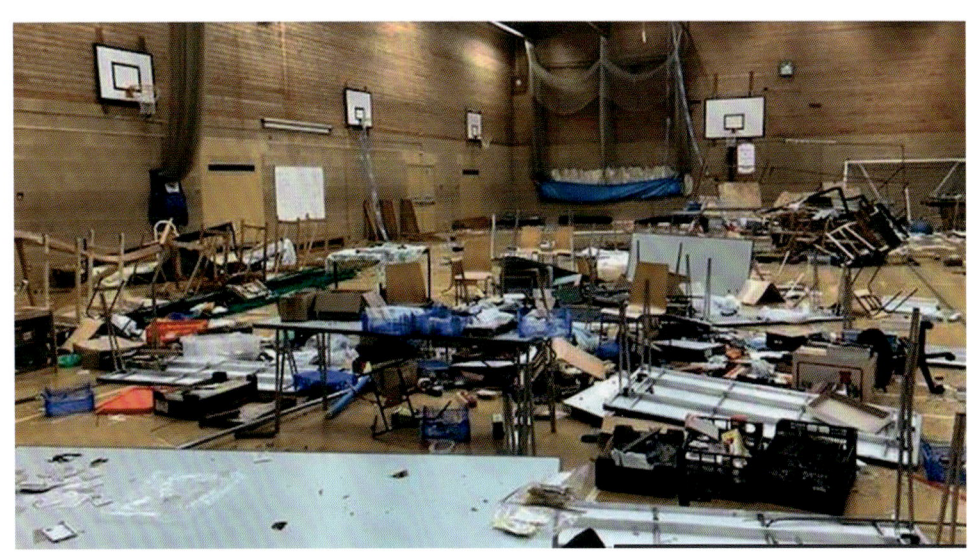

The shocking aftermath of the Club main display located in the school sports hall that fateful Saturday morning. You can make out parts of layouts and displays, but it certainly does not look as expected for a public open day. (MDMRC)

The Market Deeping Club was offered a diorama of the front environs of London's famed Euston station pre-1880, going as far south as the Euston Road. It was traditionally constructed, with a wood and hardboard carcase surfaced by cereal packets and matchsticks. In some areas we have retained this with new detailing, in others a detailed reskinning has been undertaken. In a new-build expansion, doubling the board size, we have used 3D printing and lighter-weight materials extensively, to add more detail to the scratch-built model.

It is not a precise architect's model. We have limited the board to the older eastern part of the station and simplified some of the trackwork. Some of the roads and buildings are a 'best fit' to ensure they can be incorporated into the available space.

Part of the build process has involved a deeper than normal dive by the Club into the history of the prototype London Euston. A date had to be set for which we could get rolling stock and which post-dated the modelling of the Euston Square parkland, but pre-dated the blocking of the vista by the joining of the Victoria and Euston hotels. We settled on 2 July 1875, as we know that at that time the Sultan of Zanzibar was heading north for a Birmingham trade visit and with a degree of pomp. Unfortunately, his ship had originally arrived at Gravesend rather than Liverpool, so the full state greeting against a boat train arrival couldn't be performed. This date allowed us to provide information for a display stand linked to the anti-slavery actions of those years, as well as a red-carpet vignette in the station itself.

We will share in the following pages research into the growth of London, the people of Somers Town, the operation of the London and Birmingham Railway (L&BR) and latterly the London and North Western Railway (LNWR) within the station, where you had to seek out your platform, and were likely to get lost. Alongside this is the building of the model and some tips and techniques. Research and model have coexisted well during the build. Occasionally a key facet of a building has been found after completion. Only the essentials have been revisited, otherwise we would never complete the model.

The model of Euston was adopted and curated by the author on behalf of the Market Deeping Club during the Covid lockdown. It was during this period of lockdown and social isolation that much detailing and expansion took place. Above all else it has been a great experience to research and build elements of the station – 3D printing was extensively used, enabling reproduction en masse of chimney pots, window frames, ridge tiles and detailing pieces to literally hang on the skin of the model. Taking photographs and then colour laser printing these as brick sheets and roof tiles was also utilised.

The old Euston station can be seen as a confusing mass of buildings and platforms that grew organically, in the same manner that the railway traffic and loadings grew. This was in a reactive fashion, without any true

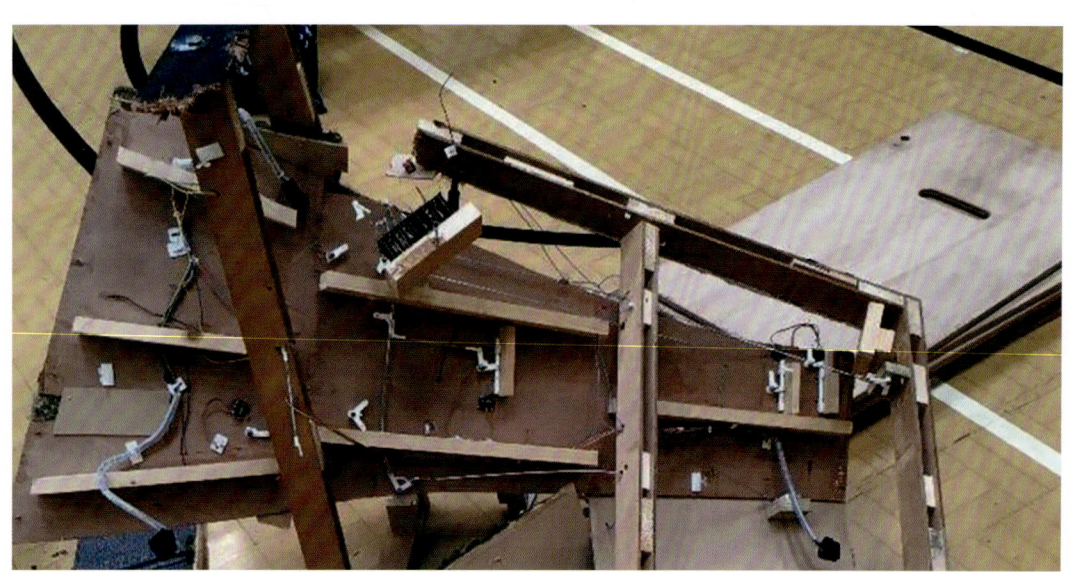

The detail of vandalism: one of the layouts destroyed. Portable exhibition layouts are normally built for rigidity, yet light and portable. Unfortunately, this makes them vulnerable to damage. (*MDMRC*)

contiguous strategic plan. It became the butt of contemporary jokes, perhaps culminating in the 1892 lyrics by George Le Brunn for Marie Lloyd in the music hall:

Oh! Mr Porter, what shall I do?
I want to go to Birmingham,
And they're taking me on to Crewe,
Take me back to London, as quickly as you can.

With hindsight some great architecture has been lost, but the negative reputation plus the need to modernise the railway resulted in the inevitable 1963 loss. It breathes again today in miniature form.

Above all else we are looking to deliver a holistic book that gives a wider view of the lost Euston, why was it there, how it grew and evolved. The model has given a physical framework, and the research a reasoned context. As a team we love to chat at shows, and the materials are used to give more depth to the model.

The author is perhaps a good example of a hybrid modeller: teenage attempts at baseboards for OO, which were bulky and could not move from the bedroom, resulting in eventual demolition and boxing of assets; family life where 1:1 scale trains were played with for many happy years at a museum MPD site in Southall West London, and everything else lived in the loft; becoming a part-time trader to sell die-cast and kits at model shows and keep

The challenge of portraying the spirit of the 1870s in 4mm to the foot. Horse-drawn vehicles, near pristine Georgian and Victorian grand buildings, a bustling station and the landmark vista northwards up Euston Grove. This view was soon to be lost when the hotels joined to become a single edifice with an access tunnel beneath. (*Author*)

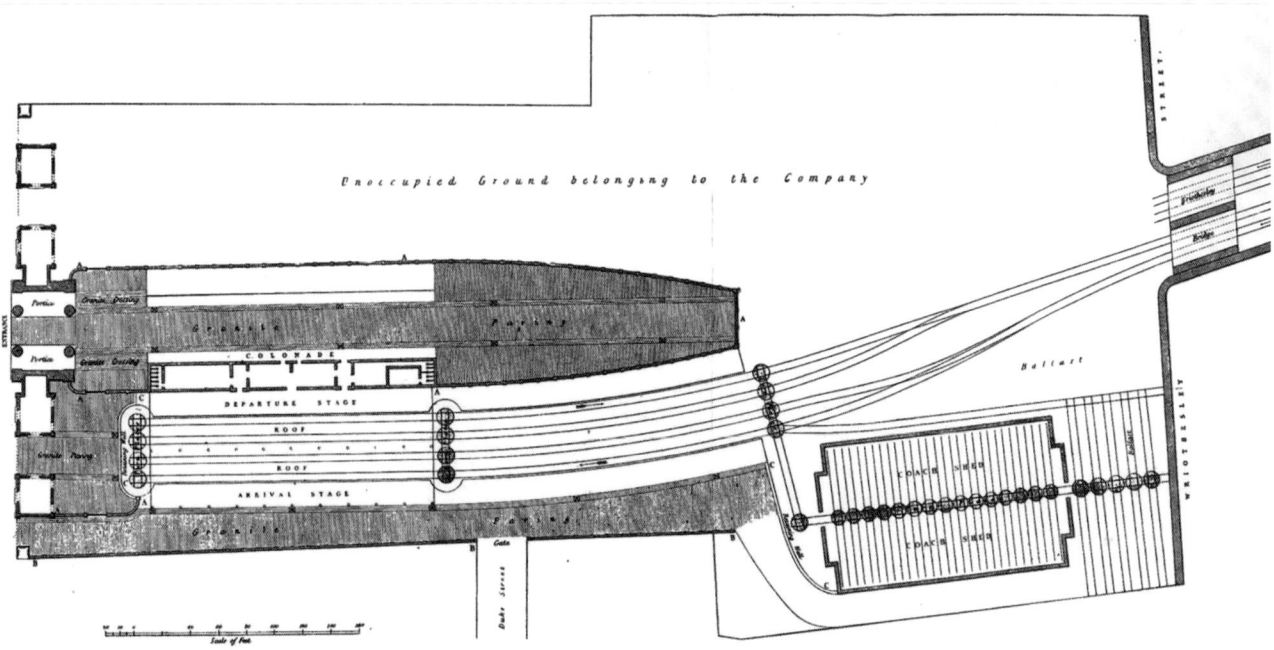

Euston Grove station. The plan of the 1838 station in original form. The monumental entrance provided by the Doric Euston Arch and lodges was symmetrical to the first station. As it grew over time this sense of proportion and purpose was slowly lost. (*Frederick Walter Simms, 1838*)

View of the current electrified Euston site taken from a drone in early 2022, with the key elements of the original station located by the author. While there was outcry at the demolition to build the modern Euston, the London, Midland and Scottish Railway (LMSR) was also aiming at a wholesale rebuild project of what was admittedly a confusing and worn-out station by the 1930s. This was interrupted by the Second World War, after which there was deemed to be no budget available. Interestingly, a similar 1900 rebuild plan has been located during the research for this book, so the Arch and Great Hall could have been lost sixty years earlier than in reality. (*Middleton Mann/HS2 Ltd, annotated by the author*)

Euston in 1936 from an Aerofilms two-seater open aircraft flying over Stanhope Street and Hampstead Street, viewing from the west over the station. From the standpoint of the Market Deeping Euston model, the cutaway is beyond the zig-zag roofing of the arrivals station in the foreground. In OO scale a full model would require a static location, whereas we wished to have a portable diorama. The target was from the Railway Clearing House (RCH) seen top left against the lines at the canopy end and incorporating more buildings off to the far right beyond the Euston Arch. Railway modelling is often full of compromises. One decision made was not to model the overall station canopy. Unlike the lofty architectural wonders of King's Cross and St Pancras, the east station of Euston had a low-lying linear canopy made of fine ironwork, which would obscure any rolling stock rather than enhance it. The station had a reputation for being smoky and dim with difficult sightlines for photographers and trainspotters. Appendix F contains an aerial image from ten years later showing the bomb damage that was sustained within this area. In comparison with the colour drone image on page 12, the whole triangle of the foreground is now becoming the new HS2 station. (*Provided under licence: image EPW049741, copyright Historic England*)

the dream alive (accompanied by his wife who by osmosis now knows a frightening amount of railway facts); eventually becoming a member of the Market Deeping Model Railway Club as retirement approached, choosing O gauge for portable and garden layouts. The saved OO pieces from childhood still exist to practise techniques and build cameos. A robust Hornby Dublo extended set serves to entertain and gently initiate grandchildren into the hobby.

Books such as this cannot exist in pure isolation. They serve as a launch pad for greater things. Read periodical magazines. Use the library service. Use the internet. Visit model shows and ask questions and see what can be done. Model shop owners and exhibitors get lonely, they love to talk! Above all enjoy yourself in discovering the pleasure of a perfect little world where the trains will always run on time.

Two views of the completed model on display on church pews, therefore obviating the need for legs. When compared to the maps and aerial photographs in this book, the challenge can be seen. Stretching from the Euston Road in the south to Ampthill Square Bridge No. 1 in the north, it is not a traditional model in that there is a non-railway element. It evokes childhood memories of a large model of the Great Fire of London in the London Museum, when it was in Kensington Palace. When anyone asks the question 'How did so much get built?' the truthful answer is plan well, take small achievable steps to keep the morale up, try to do any element once only, and for the sake of personal sanity don't keep too detailed a track of the time spent. (*Author*)

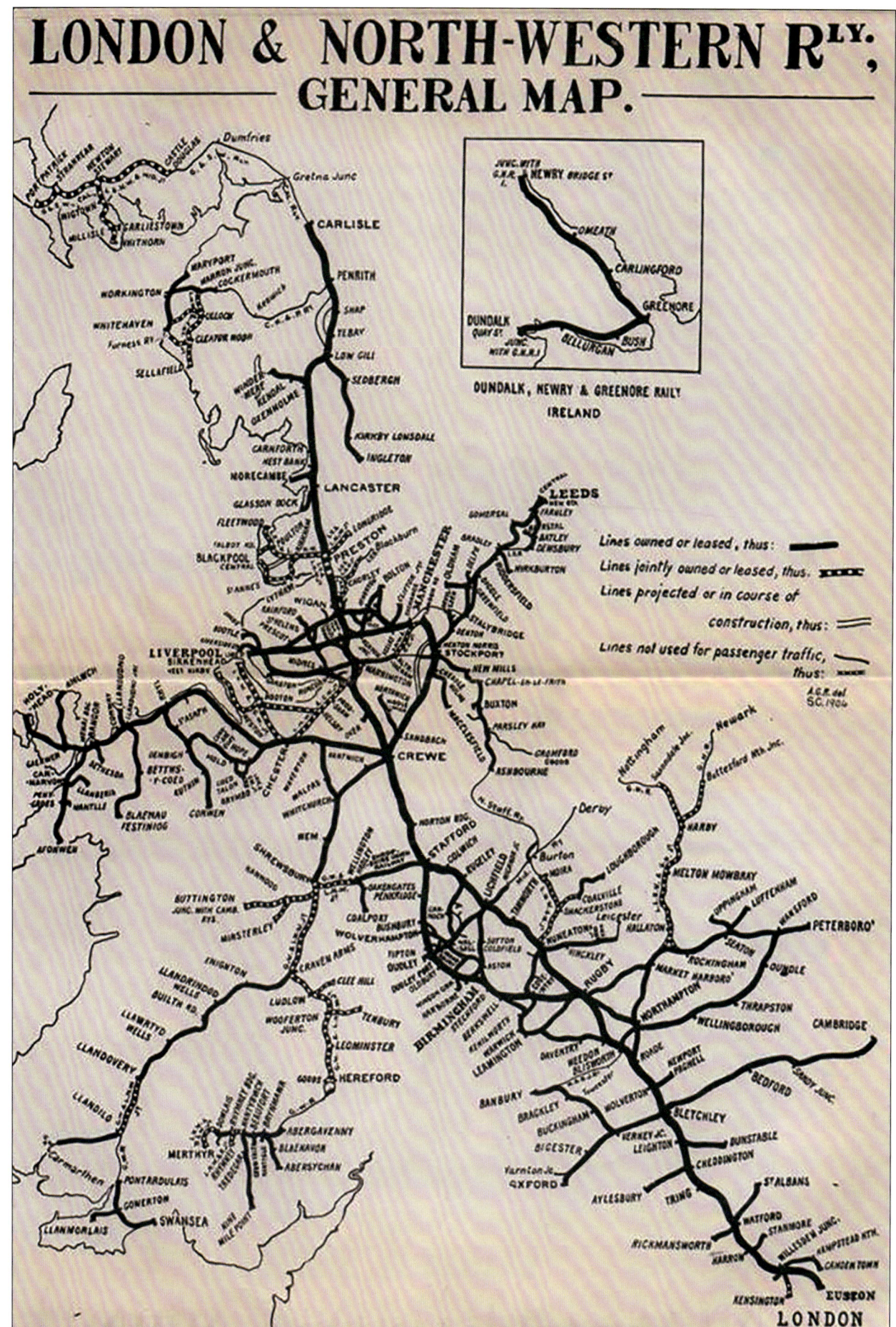

The 'Premier Line': the reach of the LNWR from Euston as published in 1908. (*Author's collection*)

A Donation to the Club

Victorian Euston station in 4mm to the foot is not something a club of our size would ever really think of. It can be regarded as a project of such a magnitude that a team would perhaps contemplate, but never get beyond a pipe dream.

Following our Club show misfortune, a generous offer was made. Were we interested in a basic diorama and boards that had been constructed in a damp relic-filled basement of an old Victorian villa-style terrace. These needed to be moved to make way for a model of Roman Verulamium as a next challenge. If we could shift it, it was ours.

The diorama itself was of traditional construction, however, the potential for public display utilising modern 3D printing, and eventually an expansion to the operating side of the station, was foreseen. All the elements were picked up and transported with one eye constantly open for parking enforcement, and the loss of quite a few brittle drinking-straw chimneys!

To fit into the available space, and given this was not a true architectural model, perspective was attenuated, streets shortened and buildings emphasised vertically to give the feeling of distance when looking along the length of a scene. It was designed for the long view, the Euston Grove vista, the feeling of Victorian London.

The challenge has been to take this core of Euston front of house, utilise modern techniques to detail it, then source accessories and vehicles. We have introduced vignettes within the model to draw the eye and feed the imagination. The park is now detailed, the hotels and the villas reskinned, the station edifice and complex expanded. We now include the Great Hall and shareholders' meeting rooms, and rails up to the Railway Clearing House at the start of the Camden Incline.

Every picture tells a story. This is just prior to dismantling. Connecting bolts had rusted shut in the damp and many generations of spiders were un-homed. The author and his intrepid wife took the model from the basement, up a slippery stairwell, through a passage, running the gauntlet of an active wasp's nest, to fill the car up to the brim and run north with a creaking manifest on board. Drinking-straw chimneys filled the rear-view mirror. (*Author*)

2
Why Was Euston Built?

London Bridge as seen from a Cannon Street train in 1956. The city and the Pool of London may have suffered Blitz and clearances but much remains. A view such as this would be familiar to a commuter of 1875, the target date of the Euston model. Wharfage, docks, cranes, barges, transhipments, trade. Accompany this with smoke, smells, pollution, noise and the old metropolis comes to life. The Thames, once the sole lifeblood of the city, was beginning to have a rival as additional arteries of steel rails were added. (*Gordon Farrow*)

Whether pertaining to a historical treatise or aiding research for a model, the broad foundation of the 'why' is important for completeness and accuracy. In this case, why did the L&BR, and its terminus of Euston station, come into being at that specific time and place?

The northern fringes of London were growing swiftly in the early 1800s. Farmland, market gardens and orchards were being consumed within a semi-planned surging change that inexorably merged the old villages. There was much linear growth along the existing road

St Paul's in 1956 as seen from Bankside and Emerson Street in Southwark. A low-rise city above which church towers and steeples still showed prominently, as in Victorian days. Warehouses dominate, their bricks dulled by layers of grime. Effectively, coal was king for major transport as well as domestic purposes as the city expanded rapidly outside the Roman walls, and over the surrounding hills. (*Gordon Farrow*)

lines. This was followed by infill responding to population pressures, as respective landowners were won over by the profits to be made from the approaches of property speculators.

To the other side of the Thames the land and property costs were lower than the city, so the embryonic railway companies serving the South were able to push through the cheaper lands of Southwark right up to the city boundary and in some cases directly into the soft underbelly. A girdle of stations had close access, with Waterloo (1848), London Bridge (1849), Charing Cross (1864) and Cannon Street (1866) breaking their way through slums, burial yards and pleasure grounds.

To the north, a hard stop was enforced partly by escalating costs and challenging geophysical terrain. The line chosen was that of the east to west: the New Road (later to become the Euston Road), Euston (Grove) (1837), King's Cross (1852) and St Pancras (1868) were all established broadly on that axis.

Prior to the steam revolution in Britain, there were already isolated rudimentary small stone and iron wagon-ways to enable horse-drawn wagons of coal and iron to move faster, and with more economy. The first known was at Prescot, near Liverpool, half a mile length, as early as 1594. This concept was enabled for widespread use at the turn of the nineteenth century by the casting of plain wheels and flanged rails by the newly-established 'iron masters' creating the 'plateways'.

Within London, the Surrey Iron Railway was opened in 1806. It was on a common user basis where the infrastructure could be rented by private horse-drawn wagons. The line ran from the growing industrial town of Croydon, heading north to the Thames at Wandsworth. There, barges could port goods into London or onto larger ships for international trade.

The Observer in September 1830 reported that the passage of goods from Liverpool to Manchester took more time than by ship from New York to Liverpool, such was the state of the roads and congestion on canals. The scene was thus set with a combination of factors around the country facilitated by growth, demand, profit and technology. Thus, the railways, as we recognise them today, were founded.

The building of canals and navigations within Britain had continued apace as the initial railways

London – dirty, noisy, sprawling, home to over 4 million souls in our model's target of the mid-1870s. The vista south-west from St Paul's is as pictured on 25 April 1955. The lack of visible Blitz bomb damage in this view means that we could reel this image back eighty years without too much imagination. (*Online Transport Archive AND-M245-8*)

Map by G. Jones of Ave Maria Lane from 1815. The New Road bypass from Paddington to Pentonville can be seen cutting east to west. Somers Town was just being developed, and The Regent's Park newly delimited. The station was constructed approximately where the wording for Somers Town shows. The site built over a new spur road for Euston Grove and eventually clipped the corner off the St James's burial ground, which at that time was regarded as full and was being decommissioned. To the east, Henry Penton's Pentonville, established in the 1870s, is largely completed. (*New York Public Library*)

Built in Bridgnorth, Shropshire, by Richard Trevithick, the 'Catch Me Who Can' of 1808 brought the concept of steam propulsion to the London urban population. The circus site is thought to have been located where the University College London Chadwick building now stands, on the as then undeveloped Bloomsbury Estate. This was just south of the New Road (Euston Road) and the Euston Square allotment gardens. The locomotive pulling a single carriage had a fare of one shilling to experience the then exhilarating speed of 12mph. Problems with soft ground led to the use of timber baulks to prevent rail breakage. However, after two months a derailment combined with reduced attendance figures led to abandonment and bankruptcy.

Above left: Trevithick's steam propulsion was built upon emerging technology and sophistication promoted by a number of prior inventors. A Newcomen condensing engine of the type used between 1720 and 1753 as seen at the London Science Museum. (*Author*)

Above middle: Early boilers were of bespoke size and construction such as this haystack boiler from Basset Pit, Derby 1796. (*Author*)

Above right: Originally nicknamed 'Beelzebub' due its misbehaviour and latterly 'Old Bess', this is a 1777 beam engine from Boulton and Watt, used at the Soho mint works in Handsworth, Birmingham until 1848. (*Author*)

Replica of Stephenson's *Rocket* seen at Shildon in 2022. The success of the 1828 Rainhill Trials was just twenty years after Trevithick's venture. The emergence of the Liverpool and Manchester, and Stockton to Darlington railways, triggered speculation further south. Freight was thought to be the primary traffic, major loadings of passengers were an unplanned-for bonus. (*Author*)

were founded – investments still came in and the system was in heavy use. As the national focus changed to investment in railways, they consequently provided an ideal source of engineers with a knowledge of public works, plus proven techniques of how to minimise gradients through the building of embankments, tunnels and cuttings. There was also a ready-to-use work force that could easily be transferred over.

The establishment of the Liverpool and Manchester Railway Company was based upon the richness of trade from the docks of Liverpool and the need for raw materials to reach Manchester and finished goods to travel outwards. Proposed returns to investment resulted in the first boom of 'Railway Mania'.

Upon creation as a company in 1833, the L&BR sought a best return for investment in planning its line between the two cities. The initial prospectus to investors offered the following:

> First, the opening of new and distant sources of supply of provisions to the metropolis; second, easy, cheap and expeditious travelling; third, the rapid and economical interchange of the great articles of consumption and of commerce, both internal and external; and lastly, the connexion by railways, of London with Liverpool, the rich pastures of the centre of England, and the greatest manufacturing districts; and, through the port of Liverpool, to afford a most expeditious communication with Ireland.

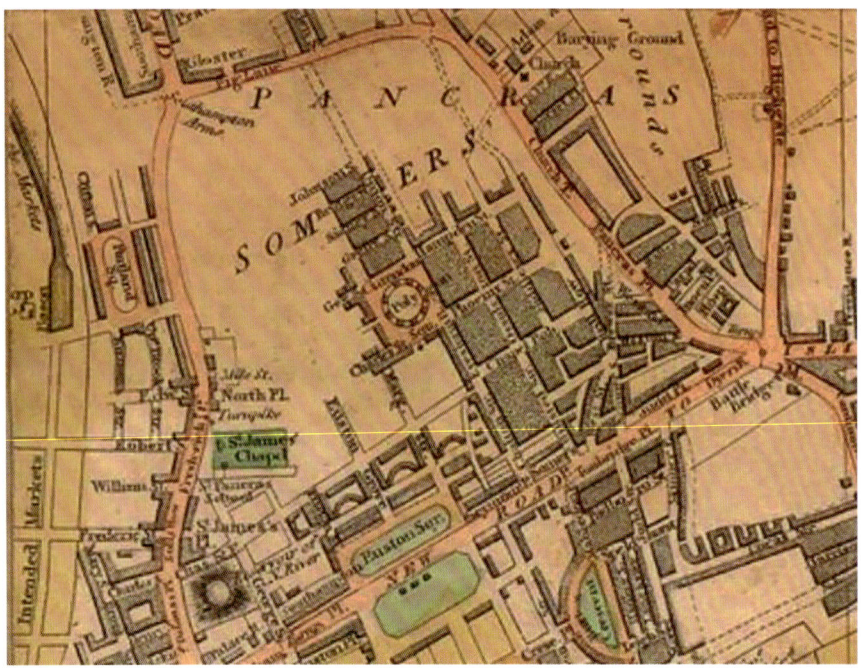

Map by Laurie and Whittle, 1817. This demonstrates the cutting of Euston Square gardens by the New Road and the prospective building along Euston Grove to establish the station frontage road of Drummond Street. The Polygon in Somers Town can be seen in the top right quarter of the map as it became encapsulated in Clarendon Square. Below Euston Square the Bloomsbury development for the 5th Duke of Bedford by Thomas Cubitt was bringing higher-class housing up towards the station environs. As he began to make money from publications, Charles Dickens was to inhabit a dwelling on Tavistock Square, not far from his childhood home in the Clarendon Square Polygon. (*New York Public Library*)

The original proposal was for the southern terminus to have been between Chalk Farm and Camden, and no further into the growing metropolis. The primary challenge being the steep fall in gradient approaching the Thames as the line passed onto river gravel terraces. The pace of technological change in the railway industry was to swiftly overcome these issues and so a terminus position closer to the city was sought.

The most appropriate target was purchase of land owned by the Duke of Grafton of Euston Hall in Suffolk, to the west side of the growing Somers Town housing developments and to the east of newly-established Regent's Park. Although a few demolitions were required, this natural valley proved to be an ideal location. So far as an Act of Parliament was concerned, this was, and still is, a branch line from Chalk Farm.

The L&BR appointed George Stephenson (1781–1848), often called the 'Father of Railways' and regarded by many as the leading engineer of his time. He was thought to be illiterate until age 18 and had a challenging thick Northumbrian dialect that his secretaries struggled to keep up with. This can be witnessed by the number of corrections to dictation in an 1822 planning book of his for the Stockton and Darlington Railway, recently rediscovered by Network Rail in the National Railway Museum. At one time Stephenson proposed Marble Arch as the line's end.

As with today's HS2, contracts for the modern high-speed railway, landowner negotiations, plus an Act of Parliament to allow compulsory purchase, were accompanied by sale for tender of contracts to engineering companies. The L&BR was content to keep an oversight, to maintain standards and quality of execution of works.

The physical groundworks for the L&BR took place at Chalk Farm on 1 June 1834. As the name implies a geological outlier of easier to work rock existed there, and cuttings created contributed a good material for embankments. However, associated with the chalk were also significant London clay strata trapping water. This the later London Underground railway tunnellers would find exasperating. As a result of underestimation of the physical conditions, the contractor for the Primrose Hill cutting and tunnel section promptly went bust, and the L&BR had to take over directly.

While the overall desire of the L&BR was to reach Central London, the costs involved of purchase and demolition precluded such a location, as did the ambivalence and sometimes hostility from the city. This response was echoed by the Midland at St Pancras and the Great Northern at King's Cross, all having to stop at the same roadway in their quest for suitable termini sites. In the south of London, the lines coming up through the cheaper suburbs to establish termini at Waterloo, Victoria, Charing Cross, London Bridge etc. were closer to the city.

3
Railway Development and Euston

Grand Junction Railway 2-2-2 locomotive No.1868 *Columbine*, built at Crewe in 1845, prior to the merger with the L&BR to form the LNWR. (*Author*)

While most people think of the British railway system as a Victorian invention, the L&BR proposal and earlier northern railway builds were made during the reign of George IV, so they can be considered Georgian entrepreneurship.

London and Birmingham Railway (L&BR)

The L&BR was established in 1833 as a merger of two other schemes. It had the aim to build a railway line between London and Birmingham. This was despite the immaturity of the steam railway technology at the time. The line was completed in 1838 and was 112 miles long. It was a major engineering feat at the time and involved building tunnels, viaducts and bridges, as well as navigating challenging terrain. The line was designed to be a high-speed railway, with a maximum speed of 45mph and opened in progressive stages.

Travellers of today would be pleased to hear that the opening also included a rail replacement

The town of Shildon's rail trail has examples of the different styles of rail. Whenever there was an innovation in the locomotives and stock, the rail technology had to change as well to cope with the dynamic downthrust and axle weights. Classed by engineer Thomas Tredgold as a 'cast malleable iron edge rail', it moved through several different iterations of shape and weight into steel. The initial Euston setup was a station with granite setts in the circulation area behind the Doric Arch and lightweight rails, since the Camden Incline was cable-worked into the station.

stagecoach service. Nothing changes! Initial service consisted of just three 'up' and three 'down' trains a day.

The L&BR was highly successful and profitable, carrying both passengers and goods between the two cities. It was also the first railway company to pay a dividend to its shareholders, which encouraged investment in the railway industry.

Euston Grove Station

The Euston Arch was a monumental gateway to the railway industry designed by architect Philip Hardwick between 1837 and 1838 as part what was originally named Euston Grove station. Celebrating London's first intercity station, this propylaeum of the Doric order stemmed from Hardwick's tour of Roman Italy in 1818–1819.

This classical Roman architecture consisted of a grand central arch flanked by smaller lodges on either side, creating a visually impressive entrance to the station. The structure was made of stone from Bramley Fall in Yorkshire and stood at a height of approximately 21m (70ft) with a width of 20m (66ft) and a depth of 13m (44ft). The overall cost at the time was £35,000 with William Cubitt as builder. Not everyone was a fan, Augustus Pugin, a Gothic proponent, called it 'a Brobdingnagian absurdity'.

The Euston Arch served as a symbol of the Victorian era's engineering prowess and the

The original style of entrance was imposing and uncluttered, as the main station building was a two-storey lateral construction behind the eastern side of arch and lodge. The naming of the station as Euston Grove was due to the previous land ownership by the Duke of Grafton of Euston Hall in Norfolk. (*LNWR postcard, Author's collection*)

The model Arch and lodges removed from main diorama context. (*Author*)

The original Euston station. A simple lateral station building behind the lodges and Doric Arch façade, and a lightweight stressed iron canopy designed by Sir Charles Fox, which was a technical innovation for this time. Just two wooden stages or platforms were required. Shunting was initially by hand, pinch-bar or horse with numerous small turntables. Some smaller locomotives used the westernmost lines while the eastern lines to platforms were initially rope-worked. (*LNWR postcard, Author's collection*)

growing importance of rail travel. It became an iconic landmark in London and a prominent feature of the low-rise city's skyline. The Arch was unembellished until the LNWR had the word 'Euston' inscribed and highlighted in gold leaf in the 1870s.

A prolific architect, Hardwick was also responsible for Great Hall, University College London (1827–1831), part of the University College London campus. The Great Hall was one of the earliest buildings constructed for the institution, featuring a neo-classical design with a columned portico. He also designed the warehouses and other structures for St Katharine Docks (1827–1828), New Church of St John, Bethnal Green (1826–1828), Highgate Cemetery Chapel (1839), Bayswater Cemetery Chapel (1839–1840) and the Great Western Hotel at Paddington (1851).

Railway Mania

The Railway Mania of the 1840s was a period of intense speculation and investment in railway projects in the United Kingdom. It was a time when many people believed that railways would transform the country and bring about economic growth and prosperity. The desire to generate wealth through share investment had been through several boom-and-bust cycles, perhaps the most infamous of which was the South Sea Bubble of just over 100 years prior.

In the 1840s, the construction of railways was a new and exciting industry, and there was a great deal of optimism about the potential profits that could be made. People invested heavily in railway companies, and the stock prices of these companies skyrocketed. George Hudson, the 'Railway King', was a key figure of this time, later discredited for sharp practices.

This cycle of investment was fuelled by several factors, including the growth of industry and commerce in the UK, the popularity of rail travel and the availability of new investment opportunities. At the time, railways were seen as a revolutionary technology that would transform the way people and goods were transported.

Investors were eager to get in on the ground floor of this new industry, and many were willing to take on significant financial risks in the hope of reaping large profits. The frenzy of investment led to the creation of many new railway companies, some of which

Building the line from the original terminus at Chalk Farm to the revised end of the line at Rhodes Farm in Euston Grove. Some tricky clay seams in the London surface stratum gave engineering challenges from Hampstead right through to Camden. The line involved demolition of recently constructed dwellings, in a similar fashion to the Westway in London demolishing newly-built blocks of flats adjacent to the Great Western Railway (GWR) main line in the early 1960s. (*LNWR postcard, Author's collection*)

The construction of the Camden Winding House for powering the initial cable system into Euston. This was in use until 1844, pulling loaded trains of up to twelve coaches on the eastern lines into the main station. It was dismantled when locomotive technology gave better tractive effort alongside the use of pilot locomotives on heavier expresses. The Grade II listed vaults and chambers still exist, and apart from being part-flooded to the same level as the Regent's Canal, they are intact with today's main line still running over the top. (*John Cooke Bourne, 1839*)

were highly speculative and had little chance of success.

Railway Mania came to an end when the bubble burst, and many of the newly-created railway companies failed. This led to significant financial losses for investors, and the UK economy suffered a recession as a result. It also allowed for a number of companies take overs and mergers to take place.

Despite the failures and financial losses, the Railway Mania of the 1840s played an important role in the development of the railway industry in the UK. Many of the railway companies that survived went on to become highly successful, and the construction of railways helped to transform the country's economy and infrastructure.

The LNWR company was formed on 16 July 1846 by the amalgamation of the Grand Junction Railway, L&BR and the Manchester and Birmingham Railway at the height of the railway boom, giving Euston direct services further north, thus increasing traffic and passenger numbers.

London and North Western Railway (LNWR)

Stimulus for the creation of this major company came in part by the GWR's competitive plans for a railway north from Oxford to Birmingham. The network at this point was approximately 350 route miles in total, connecting London with the industrial and railway centres of Birmingham, Crewe, Chester, Liverpool and Manchester.

The headquarters of the company was located at Euston railway station, which already had a growing office provision, and it was regarded as wise to keep close to the political heart.

As traffic increased expansion needs were met by the opening in 1849 of the Great Hall. The original arrival and departure platforms were deemed inadequate and rebuilt with extra faces to the track. As the station canopy requirement grew, reference was made to the new-build rival company stations nearby. The gloomy visual inadequacy of the original setup was seen by the cash careful LNWR, and the existing canopy was raised on jacks to match new-build expansion being made with the same casting style. The station fronting onto Drummond Street grew to occupy the main block between Cardington Street and Seymour Street. Further expansion resulted in two additional platforms and offices in the 1870s, and four more eastern platforms in the 1890s. The station topped out at fifteen platforms, until the 1963 rebuild.

Early on, the LNWR self-styled itself as the 'Premier Line'. Merger with the Liverpool and Manchester Railway of 1830 plus the LNWR main line linking London, Birmingham and Lancashire made it become pre-eminent, as the largest joint stock company in the United Kingdom. The LNWR collected a greater revenue than any other railway company of its era and was a founder of the Railway Clearing House coordinating other railways (see Chapter 8).

The 1840s and 1850s increased Euston's reach. The Grand Junction Railway purchased the North Union Railway in 1846, allowing the LNWR to operate as far north as Preston. Some astute leasing took advantage of the 1859 Lancaster and Preston Junction Railway amalgamating with the Lancaster and Carlisle Railway. Now the LNWR had direct access from London to Carlisle. In that same year the Chester and Holyhead Railway was brought into the fold. This opened the Irish Sea ferry routes from the Isle of Anglesey and the Irish Mail. Likewise, the boat train traffic became lucrative, connecting the port of Liverpool international steamer services to the capital.

With each merger, administrative and facility consolidation took place, and offices, management, manufacturing and maintenance facilities benefited from standards and economies of scale. By the mid-1860s locomotive construction and maintenance was concentrated at the Crewe Locomotive Works, carriage building was done at Wolverton and wagon building was concentrated at Earlestown. Euston remained the anchor as headquarters of the self-styled Premier Line and after the 1921 creation of the 'Big Four', the LMSR. In 1934, Euston House was built to the north-east of the station complex, in 1947 becoming home to the British Railways Board until the disbandment of that body in the late 1990s.

By the time of the portrayal of the Club model in 1875, London Euston was

LNWR Precursor Class 4-4-0 leaving Euston with Scotch express. (*TC35 printed by Richard Tilling, Online Transport Archive JHMC*)

directly connected with the major cities of Birmingham, Liverpool and Manchester, and (through cooperation with the Caledonian Railway) through to Edinburgh and Glasgow. Effectively today this is the WCML (West Coast Main Line). The LNWR also ran a ferry service which linked Holyhead to Greenore in County Louth. In Ireland the company also owned the Dundalk, Newry and Greenore Railway (DNGR).

The LNWR also established a number of routes to compete or prevent competition. For example, a main line connecting Liverpool and Manchester with Leeds, plus a number of cross-country secondary routes covering Derby, Nottingham, Peterborough and South Wales. The statistics of the LNWR in the early twentieth century reflect a company that was firmly in control. Route mileage exceeded 1,500 miles and there were 111,000 railway servants on the payroll. Euston had grown by this time to include new offices fronting Drummond Street (swallowing the easternmost Doric lodge) and purchasing had begun of existing buildings in the Somers Town and Camden areas for specialist functions.

One year prior to the 1923 creation of the 'Big Four' and the LMSR, the LNWR performed a major consolidation. The Lancashire and Yorkshire was merged, along with the North London Railway and Shropshire Union Railway. This gave a final route mileage of over 2,700 miles (including joint powers and leased lines).

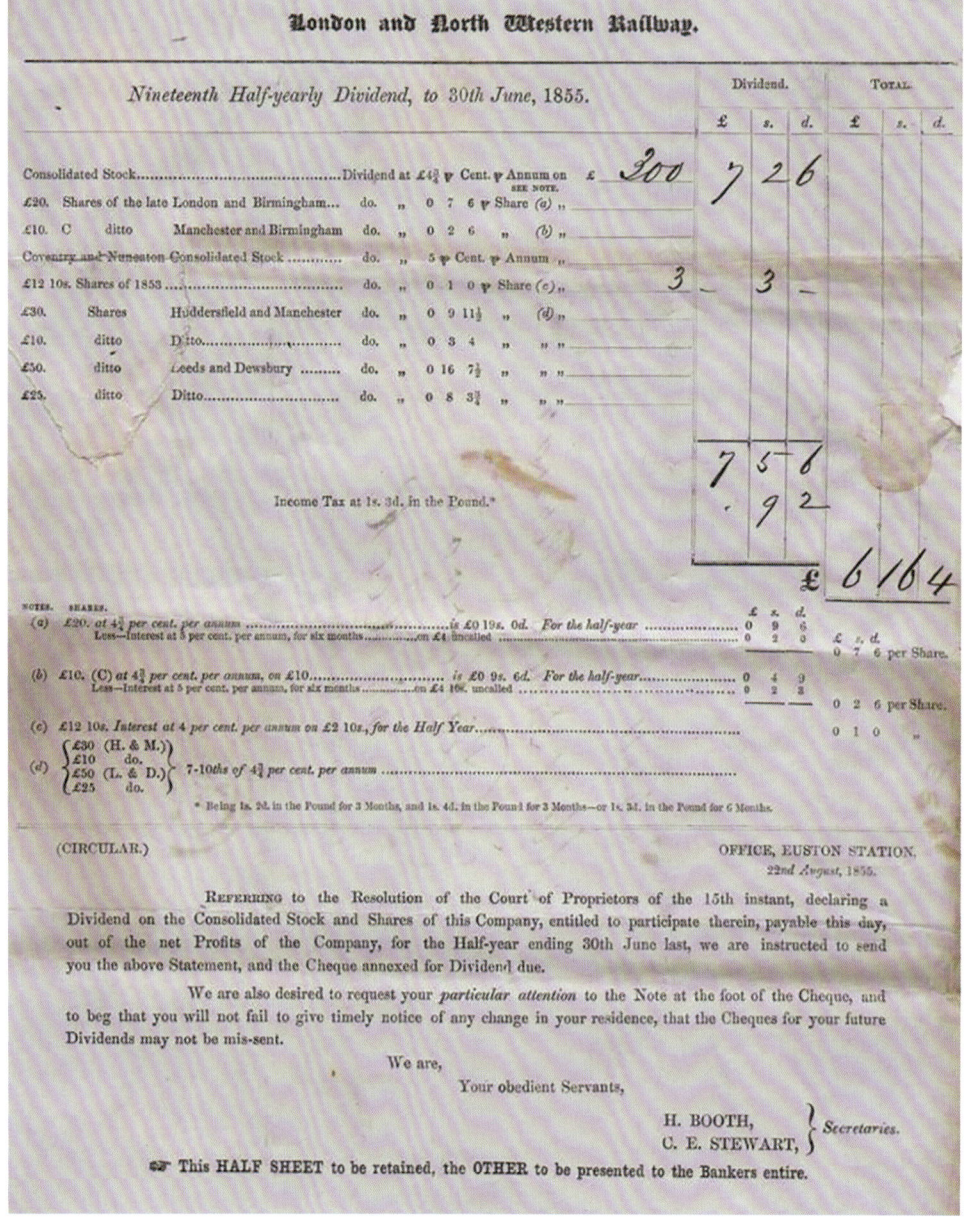

Although traffic receipts in theory covered manpower and infrastructure costs, the issue of company shares created money for investment purposes. This cash pot could cover new technology, new routes, infrastructural improvements or buyout of rivals. Interestingly the wayward nature of railway company accounting and declaration brought forth many of today's standardised accounting practices A British public quoted company also needed to issue regular dividends to investors in order to remain attractive and keep its reputation. This was especially important during the Railway Mania years, an economic cycle that occurred several times in the Victorian era. This was when many independent projects were vying for the same active monetary pool. Euston station as well as housing the boardroom and shareholders' meeting rooms, also provided offices for shareholder relationship activities. (*Author's collection*)

4
The Euston Grove Vista

The original view used as the prototype for the model. The grand view from the New Road (Euston Road) looking up to Drummond Street and the Doric Arch fronting the terminus. Even though the station was an early build on the street plan it was still forced to be asymmetrical by an offset in Euston Grove imposed by the Regency villas. Of note is the driving on whichever side of the road suited, despite the Highways Act 1835 enshrining 'keep left' in law. (*Illustrated London News, 1876*)

An oblique of the same vista taken at the Stamford Model Railway Show in 2022. In common with the lithograph, the Euston Street mews buildings are not included in the model. This is so that all the main buildings could be shown maintaining monumental form. Regency and earlier Victorian buildings had much higher ceilings to their rooms so vertical scale is important here, and is a facet rarely shown in railway models. (*Author*)

To introduce details of the area modelled, this 1936 Aerofilms view is from the south-east, looking over Somers Town and the Euston station complex. The station could be regarded as being at its zenith at this point, providing crack northern expresses, with the new LMSR Coronation and Duchess Pacific locomotives about to be introduced, as well as housing head office function for the company. The two glass light domes for ticket offices show well on either side of the Great Hall in the middle of the station complex. Ampthill Bridge has a long line of black cars with free parking. Euston was served by a garage in the curved structure mid-right on Drummond Crescent. The Hopscotch Day Nursery has been built under the trees in the parkland of Euston Square east. In the bottom right at the junction of Euston Road and Seymour Street is the Euston Square fire station, a 1902 London County Council construction which can still be seen today. (*Provided under licence, image EPW049910, copyright Historic England*)

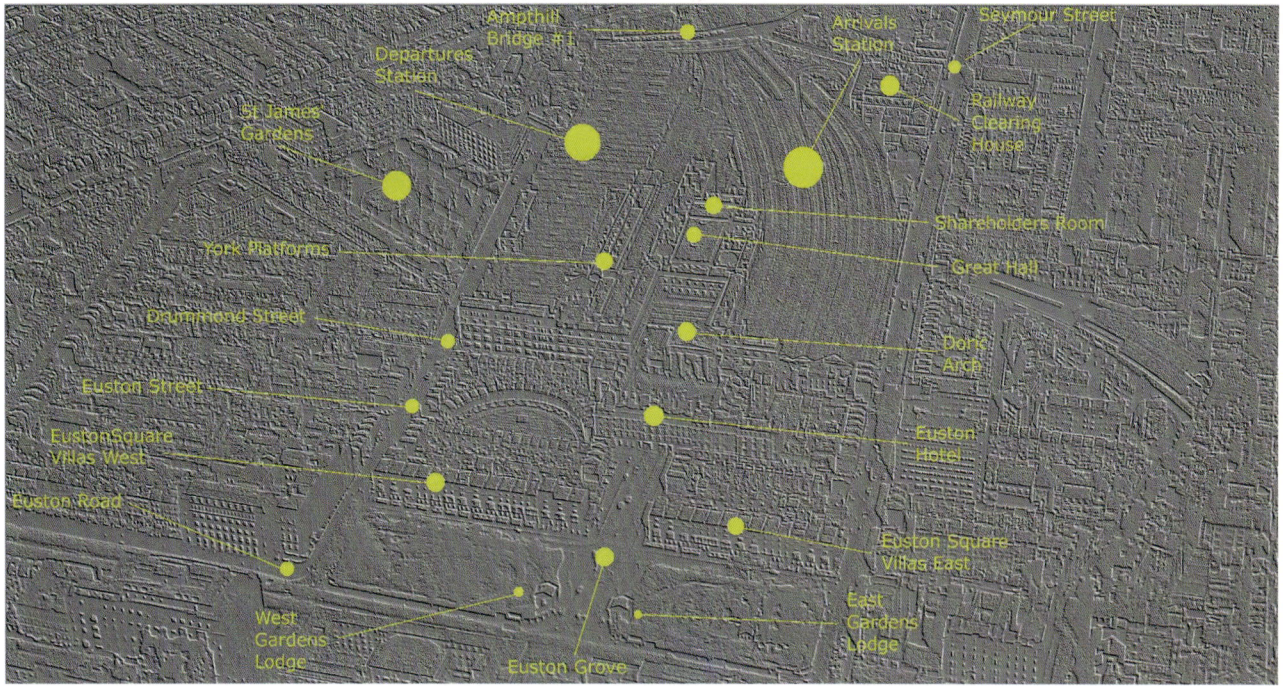

Euston Road looking north up Euston Grove towards the war memorial and hotels on 14 April 1961. The heavy wartime bombing experienced in the area has removed many of the Regency and Victorian buildings behind the garden lodges. An Eastern National service turns towards King's Cross and St Pancras with UEV845 fleet No. 367, a Bristol LS5G coach of 1953. (*Lens of Sutton, KG Carr 4955D courtesy PFidczuk*)

The same view replicated on the Club model, except in this case the coach belongs to the Lincolnshire bus company as befits a model club from that area. The Euston Road, originally the New Road, could be described as London's first bypass. It was built primarily to convey cattle to Smithfield Market. The route split the Bedford Nursery Ground market gardens into southern Endsleigh Gardens and northern Euston Square Gardens. The lodges were designed by J.S. Stainsby in 1870 and the pediment statues representing the four home countries by Joseph Pitts. (*Author*)

Euston eastern lodge in February 2023 with the major HS2 works occupying much of the image. The parkland has been utilised as a bus station and building site, losing much of the original iron railings in the process. The war memorial dates from 1921 and was constructed for the LNWR. Designed by Reginald Wynn Owen it is dedicated to the 3,719 men of the company that died under arms. Post-Second World War additional LMSR losses were added. (*Author*)

Left: Not much of the original park railings remain today and developments have progressively nibbled away at the ironwork. This section is between the Euston Road and the western lodge. In September 2020 nine activists built a tower dubbed 'Buckingham Pallets' and dug a 100ft (30.4m) tunnel in protest against the building-over for the new cab rank. (*Author*)

Right: Each lodge has a list of LNWR railway destinations inscribed on the stonework – remarkable survivors in a transformed modern landscape. (*Author*)

Right: Erwin Piscator by Edouardo Paolozzi. (*Elliott Brown CC BY 2.0*)

Below: Looking from the east to the west lodge and the Euston Tap mini pub in November 2022. The orange lines approximate the line of the now infilled pedestrian tunnel that joined the two park segments after the 1870s' remodelling. Beyond here is a 'lost' modernist statue by Edouardo Paolozzi, for now trapped within the HS2 working compound since the surrounding buildings have been demolished. (*Author*)

The tunnel portal is still extant as shown here on the station side of the east lodge looking south. The extensive cuttings leading to them shown on contemporary maps are believed to have been infilled with wartime building rubble. Certainly, enough damage was done in this area during the Blitz. (*Author*)

The tunnel and east lodge as recreated in model form. This eastern side was the grander garden, from which the first drilling works for the Paddington to Farringdon underground lines were sunk, echoed a century and half later by the protestors digging tunnels against the HS2 project. The original flat garden of the baseboard was cut out and the board dropped. The portal to the tunnel was a 3D printed object sourced online. Thus, from an N gauge tunnel mouth it was expanded in the printing software to 140 per cent, which made it suitable for OO scale pedestrians. The art of 3D printing is covered in our third modelling book, *Model Building and Super Detailing*. (*Author*)

With a portable layout one of the challenges is often the requirement to tesselate the detail faces of a three-dimensional baseboard with scenery in place. In this manner it can be safely stored. Looking here rather like the blockbuster film *Inception* are the two boards of Euston Square Gardens. They are viewed as packed away from the carrying slot of one of the endpieces. (*Author*)

The larger western park. (*Author*)

The 25in to the mile map of 1876 was used to gain the path layout and general planting of the east and west parkland, along with the cuttings to the tunnel. (*Author*)

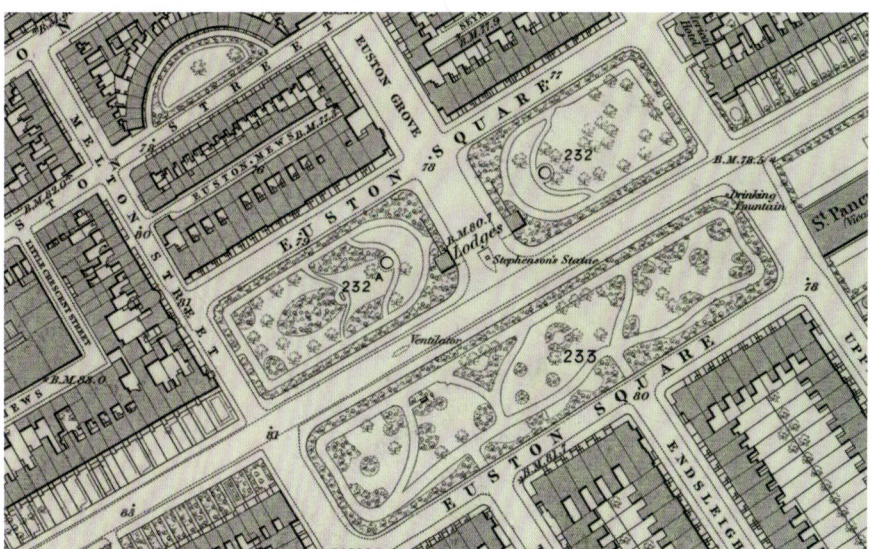

The smaller eastern park, truncated to allow for Seymour/Eversholt Street. (*Author*)

The Baseboards

The baseboards for the Euston model are a challenging mix of configurations. The Euston Square Gardens model is set upon two self-supporting boards. These can be used out of context for display at non-railway events. They store and travel inward-facing, forming a box, using end spacing boards.

The main building sections of villas, hotels and the initial frontage of the station buildings have their own triple board setup with cross-trussed braced legs underneath. It forms the central core of stability against which other boards are located and locked into place.

Where Euston deviates from the traditional railway model is that on the station side the platforms are let into the framework of a board, cutting through it. This is done because the real station was a walk-in on one-level affair from the street, but the tracks are slightly below London ground level here. Where this does assist the Club is in transportation, since all the boards are flat without accessories. What was a challenge was enabling the path of the trackbed to pass through cleanly. The actual board framework had to be

Left: Platforms 5 and 6 having just been cut out and trackbed installed. Rear face detail of the eastern office range to Platform 6 has not yet been added. (*Author*)

Below: Inter-board track connections cut into the framework for the long Platforms 5 and 6. In 1903 a kangaroo heading to Tring for the Rothchild menagerie escaped from Platform 6, and hopped over tracks to Platform 1 chased by some 100 people. (*Author*)

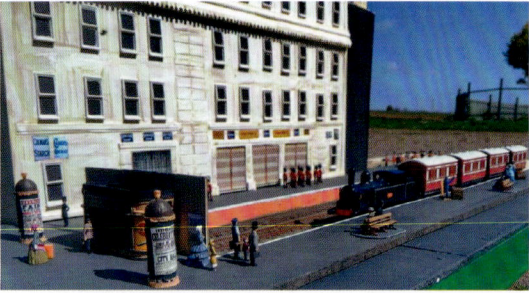

Various stages of end of station platform construction. A simple robust framework was topped with MDF and pre-trimmed to a template, to represent the physical platforms. Next, the 3mm ply trackbed was put in place, butting against the square section wooden beading, which was glued to the underside of the platforms between the framework members. Then this was topped with a cork layer and the platform sides infilled with foamboard fillets. Finally, the framework was routed down to the track bed depth and width. (*Author*)

cut and chiselled or routed out throughout four baseboards. These track boards have legs at the country end, the city end leaning on the previous board, with the central self-supporting triple boards as the stabilisation central point.

The Regency Villas (Euston Square North)

The original Regency villas of Euston Square pre-dated the 1838 station. They skew the view of the Doric Arch which ended up off centre within the overall vista. They are contemporary with the maturity of the Duke of Bedford's Bloomsbury estate to the south and follow on from the success of Henry Penton's Pentonville to the east.

The donated model villas provided the basis for moving from the equivalent of an outline drawing into a full detailed build. Because of the solid wood and hardboard box nature of the core of the building it was decided to hang 3D printed features from the outside. The alternative was to attempt a full rebuild, in which case why bother with the donated buildings at all! Apart from the extra time that would be involved with a cold start on Euston, it did not pay homage to the original modellers to not make use of their hard work.

The buildings had been constructed as mirror images of each other and foreshortened as a terrace to fit the available board space. It turned out from research that they were asymmetrical at least in the extra floor used in terrace building number two, which became the Edwards lodging hotel.

The next step was design of a standard Regency small glass panel sash window that could be mass produced to hang on the outside of the building skin, along with doors with ornate fanlights, French windows and other shapes for the upper floors. The result is pleasing to the eye when not under close scrutiny. The technique was a time saver and was continued over both original reskinned and new buildings.

An original plan was located and printed from the internet, the source of which frustratingly could not be found again. This was followed by some fieldwork walking around the Bloomsbury estate to the south of Euston. This has a number of contemporary builds that have survived. These were photographed to give clues as to depth of feature and materials.

Finally, the free treasure trove of www.thingiverse.com came up trumps with several 3D chimney pot designs to choose from, replacing the original plastic drinking straws of an earlier technology. Every room had a fireplace with a dedicated chimney flue, resulting in a wonderfully crowded roofscape.

Images such as these two Bloomsbury views assisted the interpretation of balcony construction, building ends, and the ashlar key and column design the Georgians made much use of. You find decisions easier to make if you can make field visits to an original site. If the building is no longer present, locate one constructed by a specific architect or a similar school of design. (*Author*)

The south side of Euston Square was renamed Endsleigh Gardens. It is thought this was proposed to the Vestry of St Pancras after a notorious murder in No. 4 occurred in 1877, where Matilda Hacker's body was disposed of in the coal cellar of that house and was discovered in 1879, resulting in the accusation of servant and mistress Hannah Dobbs.

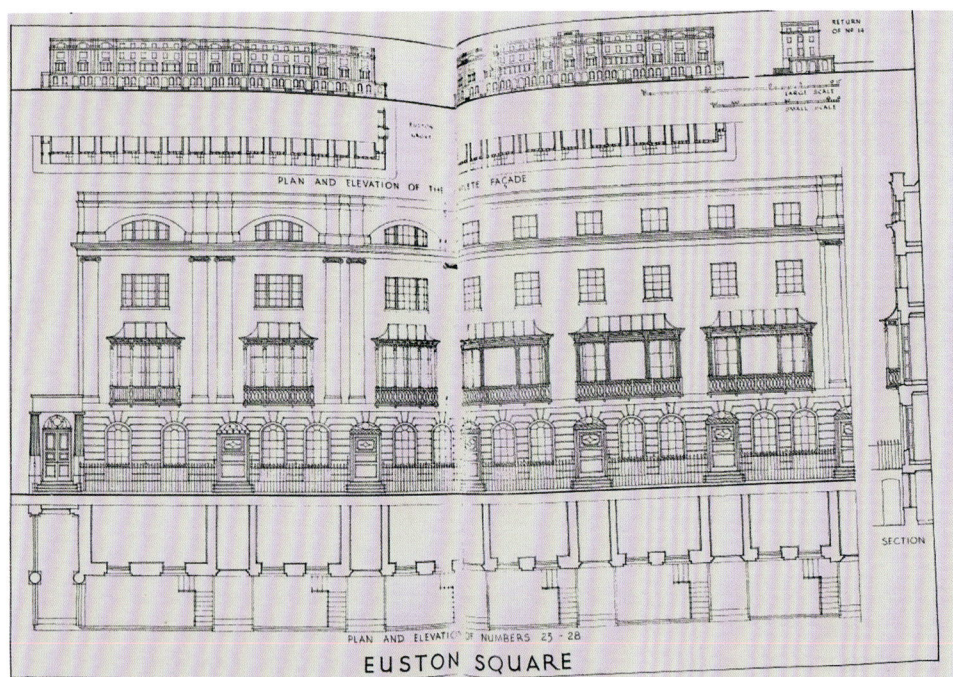

The printout of what was thought to be an original design book plan from the Euston Square villas. This greatly aided the reskinning of the original model. Finding photographs of these buildings shows that there were detail differences to this plan, especially in the number of windowpanes and the shape of upper windows. (*Source unknown*)

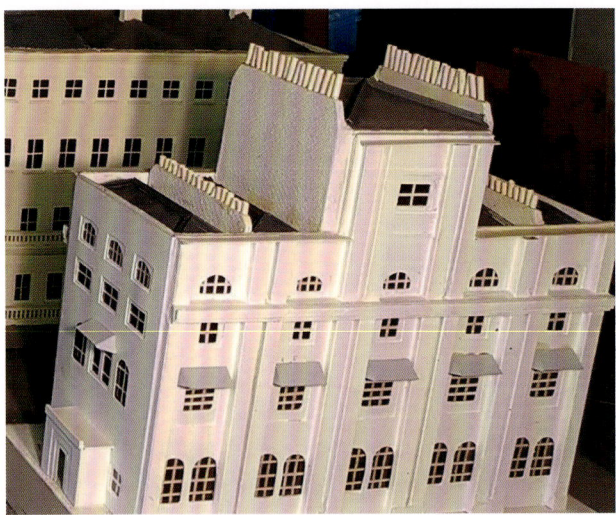

The original under-detailed building. It gives the feel of a Georgian structure but does not reward closer scrutiny. (*Author*)

Reskinning underway. Overlaying the original with 3D thermal filament printed features and plenty of biscuit box-derived card stuck on. (*Author*)

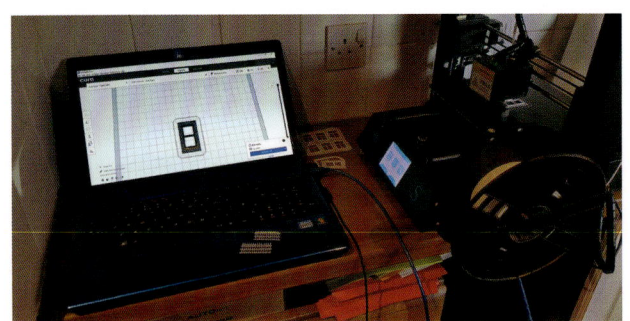

Printing out 3D sash windows using a computer-aided design (CAD) model, via a 'printer relationship' product called Curia. Fifty-four windows for the villas, and triple that number on the main station new offices. (*Author*)

Completion of the west villa. Balconies of visiting cards, fine fishing net mesh and cocktail sticks. A nice array of 3D chimney pots. Windows are hung on the outer skin rather than recessed. Field research into similar houses paid off since it felt correct when placed into context. (*Author*)

Both sets of villa buildings in position on the busy north side of the Euston Square street scene. (*Author*)

The Street Market

Apart from some early news-stall positions between the lodges next to the Doric Arch of Euston (Euston Arch), we are not aware of any street markets in the immediate locality. Most of the area in Booth's Poor Map for this part of Euston is of better classes (except sublet mews behind the larger buildings). We wanted to portray a small general market under the LNWR advertising at the end of the east villas. Some fresh produce stalls, a rags stall and a hokey pokey (ice cream) man alongside the knife sharpener. The poor would shop here and indeed sell here, hoping to make enough for the next day as times were becoming increasingly tough.

A contemporary account of a London Market by Henry Mayhew in his treatise for The *Morning Chronicle* from 1851, 'London Labour and London Poor', gives the overall sense of such a gathering that we have tried to achieve.

The pavement and the road are crowded with purchasers and street-sellers. The housewife in her thick shawl, with the market-basket on her arm, walks slowly on, stopping now to look at the stall of caps, and now to cheapen a bunch of greens. Little boys, holding three or four onions in their hand, creep between the people, wriggling their way through every interstice, and asking for custom in whining tones, as if seeking charity. Then the tumult of the thousand different cries of the eager dealers, all shouting at the top of their voices, at one and the same time, is almost bewildering.

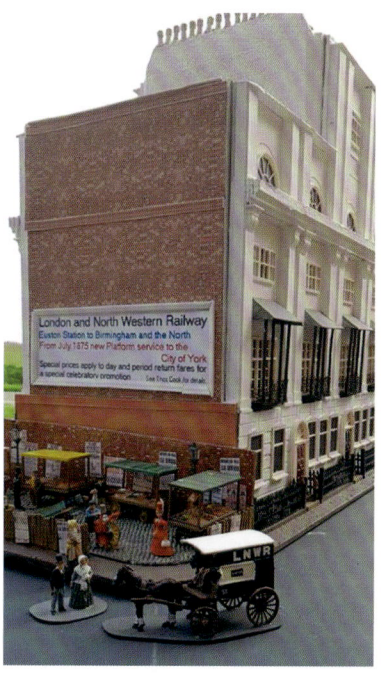

Above and right: Observation of photographs shows that Victorian London suffered from rampant illicit fly bill advertising. It appears that if it didn't move, a poster was adhered to it. Advertising varied from theatre, circus, fair, land sales, political statements, through to the final public executions. Interestingly you could have caught the Tube from Euston Square towards Newgate to see the last of these on 26 May 1868, when Irish Separatist Michael Barret was hanged for his part in an explosion outside Clerkenwell Prison. The wood fencing here is 3D printed and has been usefully rescaled for an O gauge layout as well. Plastic market barrows and cobblestones are from Wills (Peco). The chestnut/ice cream seller and knife sharpener are white metal kits from Langley Models. Euston was the first London station to link to York and Platform 9 thereafter was known as 'The York'. (*Author*)

The Pipe Yard

On Euston Street east we have located a small pipe yard for distribution of new cast iron pipes. These would have been manufactured in the nearby St Pancras Ironwork Company Foundry, which was located opposite St Pancras church. Such distribution points existed during the rapid expansion of London as the infrastructure struggled to keep up with urban developments and legislation. They were then sold off as smaller housing plots as land prices rose.

One of the triggers for much of the city being modernised was the 'Great Stink'. This was a historical event that occurred in London during the summer of 1858. The event was characterised by a pervasive and overpowering smell that emanated from the River Thames, which ran through the heart of the city. Such events had been noted during the previous ten years, but now the stench was so strong that it was said to have made people sick. Many who lived or worked near the river were forced to leave their homes or businesses. The Houses of Parliament hung lime-soaked sheets at windows in an attempt to reduce the stench.

The cause of the Great Stink was a combination of factors. At the time, London's sewer system was inadequate to handle the city's growing population, and much of the waste produced by the city's residents was dumped directly into the Thames. The river was also heavily polluted by industrial runoff and other forms of waste, which added to the already noxious odour.

The situation reached a crisis point in the summer of 1858, when a heatwave exacerbated the smell and made the stench unbearable. The situation was so bad that it prompted the government to act, and a new sewer system was designed and built to address the problem. The new system, known as the London sewerage system, was completed in 1865 and helped to dramatically improve the quality of life for Londoners. The Great Stink is now seen as a turning point in the history of sanitation and public health, and it played a key role in the development of modern sewage systems around the world.

Parallel to this was the requirement for piped fresh water for drinking. Many springs and wells in London were being polluted by human, animal and industrial waste. Dr John Snow researched the Broad Street water pump in Soho in 1854 to find it as the source of cholera. At the same time over 500 people died in the Somers Town and St Pancras areas. As Euston station and the hotels expanded so did their capability to provide clean water and divert waste into the new wastewater network.

The new London sewer system was designed and built by a civil engineer named Joseph

The pipe yard, a distribution point for cast-iron sewerage pipes at the end of the west villas. Old farm carts were often used since their open framework construction enabled temporary cradles to be nailed into place. These white metal carts are from Langley Models. The shed, fences, pipes and walls are all 3D printed and scaled to fit. (*Author*)

Bazalgette in the mid-nineteenth century. Bazalgette was tasked with finding a solution to the city's growing sewage problem, which had become a major public health issue. That solution was to create a network of underground sewers that would carry waste away from the city and out to sea. His design was based on the principle of gravity, which meant that the sewers would slope gently downwards towards the river, allowing waste to flow away naturally.

To build the sewers, Bazalgette had to overcome a number of physical challenges. One of the biggest was the fact that much of the city's existing infrastructure, such as roads, buildings and underground utilities, was in the way. To get around this, he had to build the sewers deeper underground than originally planned.

Bazalgette also had to deal with opposition from some quarters, including those who argued that the cost of the project was too high. However, he was able to convince the government to support the project, and work began in 1859.

The new sewer system was a major engineering feat, comprising more than 1,000 miles of pipes and tunnels. It was completed in 1865 and was hailed as a major triumph. The new system not only solved the sewage problem but also helped to prevent outbreaks of diseases such as cholera and typhoid, which had previously been common in the city.

As an aside, in family research after building the pipe yard, the author discovered that the 1854 cholera epidemic almost wiped out a familial branch. After losing a wife and two daughters to the disease within two months, the relative who was a pattern maker at the St Pancras iron works, went on to remarry and rebuild a family of six children. Truly a case of life imitating art.

The Victoria and Euston Hotels

Above right: The dining room of the LNWR Euston Hotel had a decidedly French flavour on 6 January 1885. By this time the budget and the upper-class hotels had merged into a single building. (*Author's collection*)

Right: London Omnibus Company Route No.1 passing along Drummond Street and looking into the end of the Euston Grove cab rank. (*Author*)

The Guests

Two sets of census returns were researched as background to the project, aiming at 1851 to reflect the early years and then the 1871 returns which equate to the model in the years ahead of hotel rebuilding.

While it may seem to indicate an unprofitable position in numbers of guests listed, on the evening of Victorian-age censuses, people were encouraged to stay home unless essential activities were involved. In this manner the assigned enumerators could then go from door to door gathering accurate information from the head of household. The numbers may have been accurate, but often the verbal delivery of a cockney or rural accent being transcribed, combined with field-based handwriting, leads to some interesting challenges.

The staff occupations of the hotels are intriguing. As expected, there are a number of porters, chambermaids, housemaids, kitchen maids, cooks and so on. Additionally there is a plate man, a carpenter, a cellar man and a scullion. In 1851 there were swill-room maids (emptying chamber pots), but by 1871 corridor lavatories and baths were being installed.

In 1851 the higher-class Euston Hotel has its own night keeper, Thomas Edwards, aged 51, with thirty-four staff present looking after forty-one guests. A number of land agents and barristers are staying, a Captain of the Royal Ordinance and a Lieutenant Colonel Brevet with his wife.

The Victoria Hotel was a slightly cheaper establishment as a sleeping-only hotel. There were twenty-five staff present that night, working under the cashier, Joseph Cornock, aged 30. The eleven guests listed staying are mainly stated as 'gentlemen' or 'merchants'.

According the 1871 census there were forty-one guests in both hotels under the care of general manager Robert Whalen, aged 61, from the Isle of Wight. In all probability he would still be managing in 1875, retirement being more of an ill-health activity. At that time there were sixty-three staff logged as being on duty or living in the properties.

By this year the concept of a census was understood and people became more detailed and forthcoming about their backgrounds. Notable guests are as follows: John Medcalf, aged 39, woollen merchant, and his wife Charlotte, aged 36; Peter G. Heyworth, aged 47, a South American merchant; Captain Bugg, aged 34 of the Artillery; F.E. Mac Nouton, aged 42, Colonel of the Horse; John Bloson, 46, ship owner; T.A.T. Swinson, 30, Captain 11th Hussars; and J.B. Bickersh, aged 44, a retired Indian civil servant.

These were transit hotels, mainly to allow arriving or departing passengers easy access to and from Euston. Smaller local hotels existed such as t he Edwards' family hotel which occupied several houses of the Regency west terrace of Euston Square in the late 1800s.

An 1880s' view on colourised postcard of the newly conjoined hotels. Both wings were designed by Philip Hardwick, opening in September 1839 on land purchased from Lord Southampton. Note the tunnel under the French-inspired central building, supported on four rows of Doric cast iron, thus providing access for Euston Grove. This extension was designed by LNWR architect J.B. Stainsby who, following the collapse of Huddersfield station canopy in 1885, was thereafter supplemented by an engineer. (*Author's collection*)

The Victoria Hotel during re-roofing and the addition of new chimneys. The dome is half a cistern float. The buildings are made from wood scraps covered in hardboard then surfaced in cereal packet card. The roof lines are all cat food sachet box corrugated cardboard, used for strength. At the time modelled both hotels had a Portland stucco finish. (*Author*)

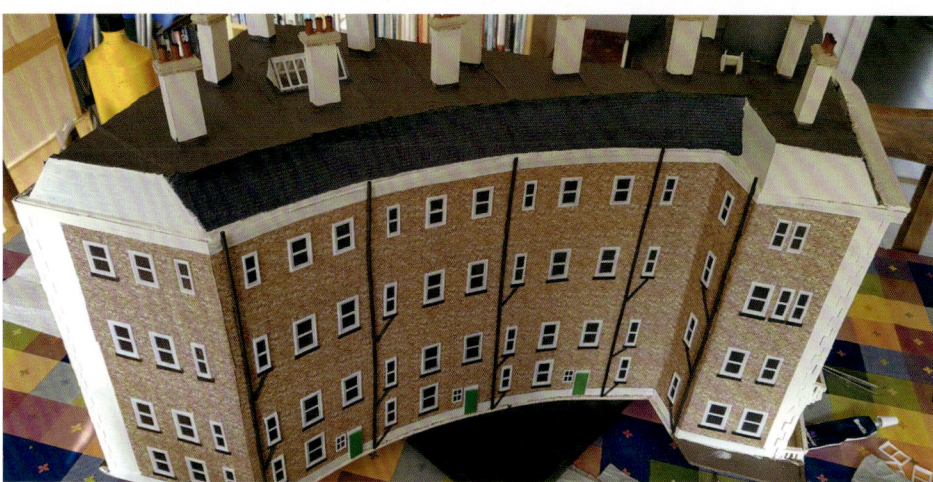

The Euston Hotel with curved rear allowing for the presence of Drummond Crescent. A new 3D printed window array was added to this side based on aerial photography. This hotel had a notable dining room with two Doric columns *in antis*. (*Author*)

The Slums

The following are extracted Police District 18 notes of the locality, taken by Ernest Aves while walking the station surroundings with police Inspector Bowles in 1898.

> Seymour Street South east corner by park. Same general character as at the north end. Third rate shops for the most part, remain pink as map. The red goes out at the SE end, and the corner itself should be purple for a common lodging house, beds 6d, 9d and 1 shilling, quiet.
>
> Seymour Street south has smaller shops, the buildings are three storeyed, and the whole of the of the rest of the street is rightly coloured pink. The coffee house at No.174 is suspected as a brothel.
>
> Drummond Street. Three and a half storeyed houses, with attics for the most part; rougher than Lancing Street. Just 3/4 persons to a house and should go from purple to light blue. Running out of it is Roberts Yard, very poor, dark blue.
>
> Euston Road near Seymour Street. Mainly shop frontages, with the houses occupied in all sorts of ways. Near here we passed the building marked on the map as Temperance Hall, with its back entrance in Weir's Passage. This place is now called Cecil Reading Club, and has a doubtful reputation for betting. In any case the quondam [*sic* 'former'] Temperance Hall has fallen on evil times.

Prostitution, gambling, illegal drinking clubs and opium dens are all located close to the main stations awaiting the unwary and the fresh-faced from the country, who would be arriving in London for the first time. The walking notes are full of such information, King's Cross having a higher concentration than Euston due to the nature of the housing stock in that area.

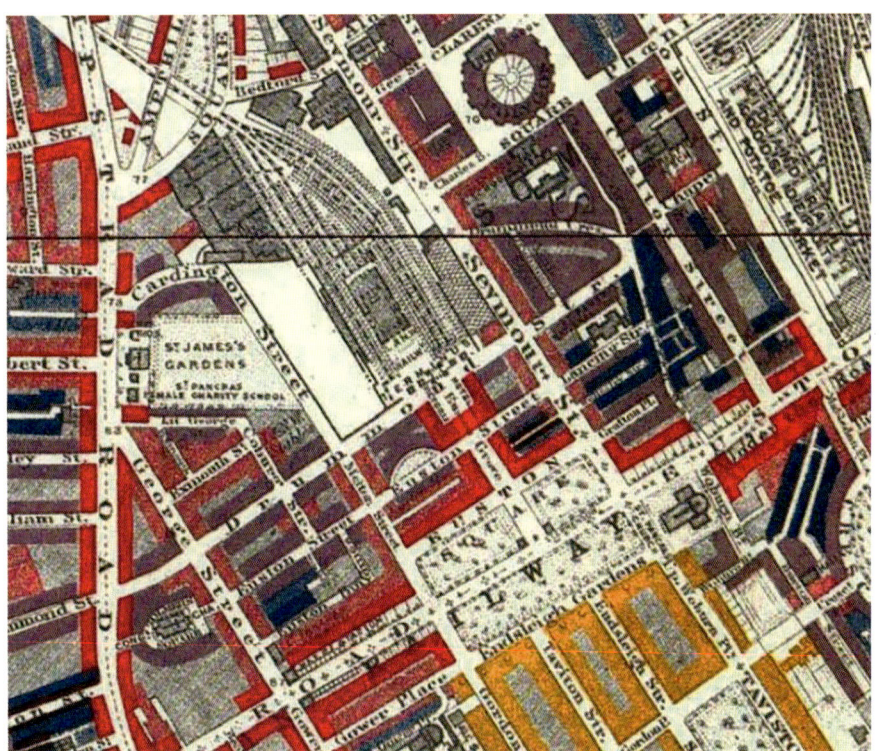

Booth's Poor Map of 1889 shows that generally the Euston area was outwardly maintaining a decent class of inhabitant. The mews behind the eastern villas was falling into multi-occupancy. High class was south of Endsleigh Gardens in Bloomsbury. As one proceeds to the north of the station The Polygon and parts of Somers Town are showing slippage as Camden came under pressure from economic change and personal deskilling of the population. Eventually many formerly grand buildings became one family to a room dwellings. The walking notes show that many homeless were in doorways and around the St James's Gardens area. (*British Library Public Domain 1.0*)

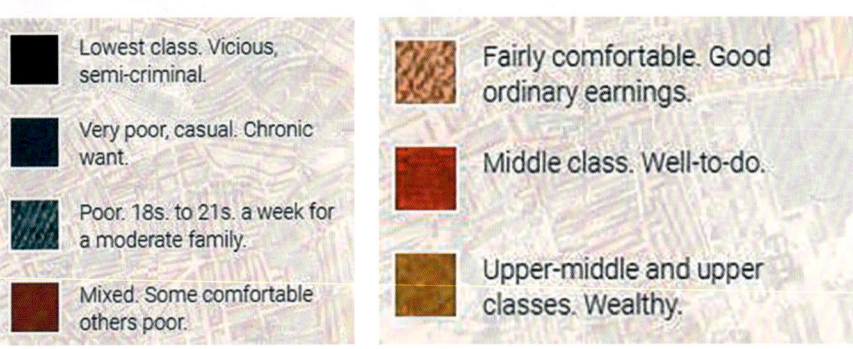

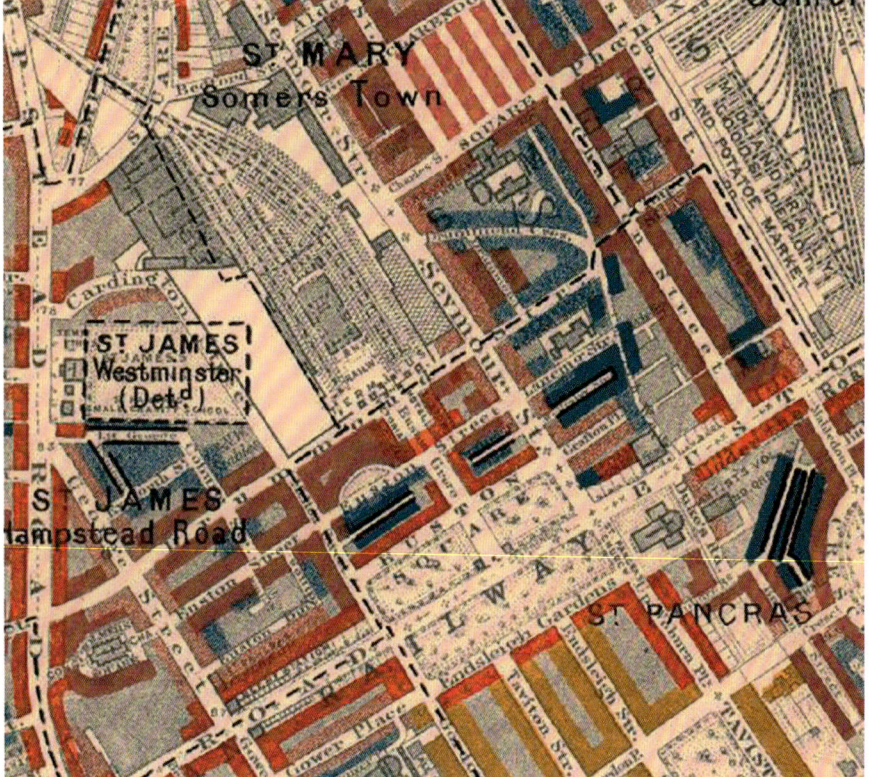

Booth's Poor Map of 1898, which is contemporary with the walking notes. Some slippages of Endsleigh Gardens in the south due to retail. The immediate environs of Euston station have held their own apart from the mews dwellings becoming multi-occupancy. Clarendon Square has now lost The Polygon, demolished as one of the worst slums of Camden and Somers Town. Drummond Crescent became run down enough to be replaced by the LNWR motor garage in the 1900s (see aerial photographs). (*British Library Public Domain 1.0*)

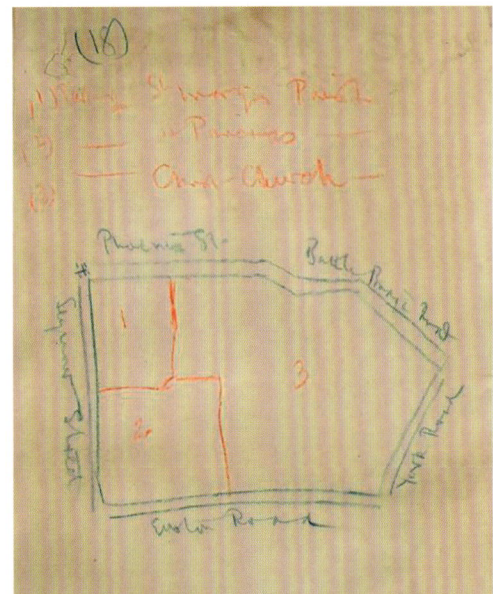

Left: Ernest Aves's notebook. Police District 18 Somers Town and Camden Town. This is the sketch map of a walk with police Inspector Bowles. (*LSE*)

Below: Example houses of South Seymour, now Eversholt, Street. The alleged brothel fell to the Blitz. (*Author*)

The Polygon in Somers Town, 1850. A circle of outward-facing houses and mews established in the late 1700s and adopted as artists' dwellings. Originally this was an isolated development surrounded by market gardens and brick clay pits. In the 1820s Charles Dickens lived at No. 17 in The Polygon. This was just after his father John returned from debtors' prison. Dickens's *Bleak House* character Harold Skimpole was based in The Polygon. After Euston station was built, extensive multi-tenancy in each house occurred, leading to overcrowding with whole families living in a single sublet room and carrying out an unskilled trade. (*Old and New London,* Edward Walford, 1878)

> There are foul ditches, open sewers, and defective drains, smelling most offensively, and generating large quantities of poisonous gases; stagnant water is found at every turn, not a drop of clean water can be obtained, all is charged to saturation with putrescent matter. Wells have been sunk on some of the premises, but they have become, in many instances, useless from organic matter soaking into them; in some of the wells the water is perfectly black and fetid. The paint on the window frames has become black from the action of sulphuretted hydrogen gas. Nearly all the inhabitants look unhealthy, the women especially complain of sickness, and want of appetite; their eyes are shrunken, and their skin shrivelled.

Charles Dickens, 'Health by Act of Parliament', *Household Words*, 10 August 1850

To complete our research into the surroundings of Euston, and especially since we were constructing higher class landmark buildings rather than standard housing, we looked into the downfall of The Polygon in the centre of Clarendon Square. This reflects in a microcosm the social pressures experienced throughout London as migration increased, population density became dangerous to health, deskilling of the local population occurred and general incomes fell.

It was built in open countryside as a fifteen-sided development with thirty-one houses and a central garden area. William Godwin and Mary Wollstonecraft (writer/philosopher), with daughter Mary Shelley (author of *Frankenstein*) lived here around 1797. (Today there is a brown commemorative plaque in Polygon Road.) Charles Dickens rented here in 1828, when multi-tenancy had started and decline had set in. The Polygon can be seen in the 1815, 1817 and 1889 maps in this book, situated to the north-east of the station. The houses were demolished in 1890 as a notorious slum.

An example from personal family research. A skilled boot maker arrived in London from Saffron Waldron just prior to 1841 and rented accommodation in Camden. By 1851 the family appears to have sublet the property as overall incomes fell, to keep rental payments going. By 1861 younger and oldest members of the family were expected to contribute to keep costs down. They are listed with associated boot-making tasks. By now the market would be crowded with similar products, reducing margins. In 1881 the family was classed as near destitute due to falling income, living in a single room and deskilled, cutting leather, sewing soles or joining shoe components for a piece rate. Trying to avoid the nightmare of the workhouse was a tale carried through several subsequent generations, to myself.

The census of 1851 indicates that in the 31 houses were 222 occupants. Many were servants or nursemaids living in the basements and garrets, giving a generally equal male to female ratio. Among the listed professions were House of Commons' clerk, French polisher, carver and guilder, wood carver, two railway clerks, a wood turner, a clerk in the Court of Chancery, a straw bonnet maker, six pianoforte makers, a foreman of a shoe warehouse, a paper hanger, a watch maker, three artists, a poet, a brush manufacturer, a book binder, a print compilator, a landscape engraver, a music tuner, a collector for a gas company, a dancing master, a jeweller, a cordwainer, a Catholic priest, a painter in watercolours, a coach maker, an Admiralty clerk, a teacher and translator, and a soda water manufacturer.

The census of 1871 gives a very different picture: in the same number of houses there were 362 inhabitants, and far more females than males inhabitants, many 'working' rather than listed as 'lady of the house'. A degradation in job skilling had taken place and a substantial number of men were listed as out of work. Subletting had resulted in families occupying a couple of rooms per building.

At this time the occupations were listed as a House of Commons' clerk (still there after twenty years but now with several sublets) two artificial florists, three seamstresses, a shopman, a laundress, a milliner, a tailor, two gas-mantle makers, a bricklayer, a chair maker, four railway porters, a ladies capmaker, a plasterer, a saddle-tree maker, a railway servant, a musician, an inkstand maker, a needlewoman, a railway van guard (Great Northern Railway – GNR), a leather carver, a glass cutter, two dressmakers, two coach painters, an organ builder, three engine fitters, an omnibus conductor, two cab drivers, a shirt maker, three commercial travellers, a coffin-maker and a midwife.

Being a railway porter could be a lucrative position through personal tips, and many would arrange to work off-shift if an American boat train was incoming. But there was a pecking order, and many remained poor at the bottom of the heap.

On many census returns the use of 'seamstress' and 'dressmaker' can be used as alternatives for 'prostitute'. Or it can mean exactly what it says, it all depends on context. In one house the head was a lady running a board and lodging house with a number of these professions present. Two houses along was a man lodging who was working for the London Missionary Society. One can build an interesting social backstory based on just a few facts.

The vista boards linked up for a test display. The underframe trussing on the triple scenic board can be seen. The original LNWR cast-iron warning sign acts as an overall anchor to the layout itself, though sheer inertia! Winnie the cat was a constant attendant of the build. (*Author*)

5
The Station Buildings

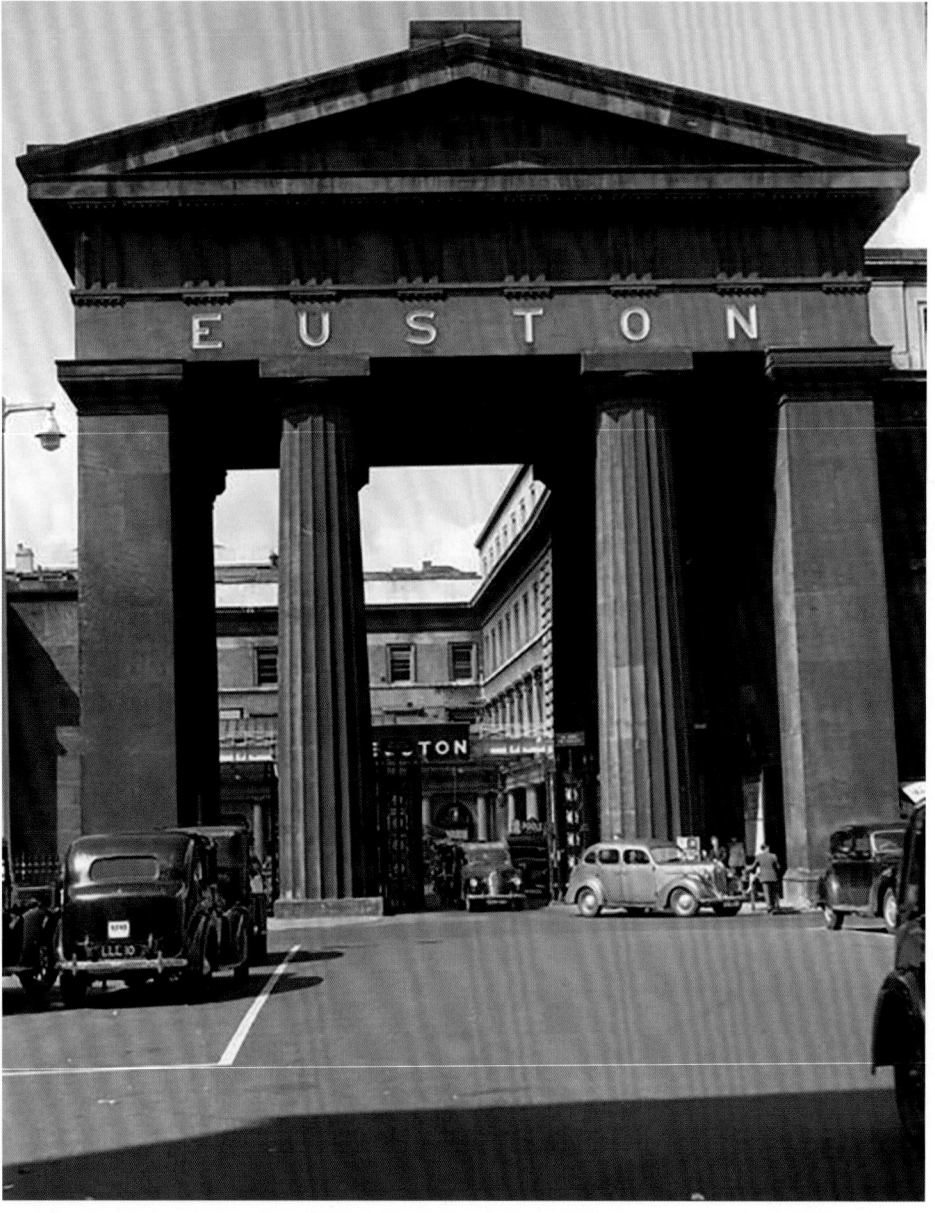

Euston Doric Propylaeum, as the scholar would term it, seen on 17 August 1953. No space was wasted – it is said there was a printshop office at the top reached by a staircase. (*Online Transport Archive AND-M137-3*)

The Chaos of a Victorian Terminus

The following extract is from *Three Men in a Boat* by Jerome K. Jerome. While it is light-hearted in nature, it hit the correct pitch so far as the rail user of that time is concerned. Stations were ill lit, smoky, dirty and lacking public address systems other than signs, chalkboards and megaphones.

> We got to Waterloo at eleven, and asked where the eleven-five started from. Of course nobody knew; nobody at Waterloo ever does know where a train is going to start from, or where a train when it does start is going to, or anything about it. The porter who took our things thought it would go from number two platform, while another porter, with whom he discussed the question, had heard a rumour that it would go from number one. The station-master, on the other hand, was convinced it would start from the local.
>
> To put an end to the matter, we went upstairs, and asked the traffic superintendent, and he told us that he had just met a man, who said he had seen it at number three platform. We went to number three platform, but the authorities there said that they rather thought that train was the Southampton express, or else the Windsor loop. But they were sure it wasn't the Kingston train, though why they were sure it wasn't they couldn't say.
>
> Then our porter said he thought that must be it on the high-level platform; said he thought he knew the train. So we went to the high-level platform, and saw the engine-driver, and asked him if he was going to Kingston. He said he couldn't say for certain of course, but that he rather thought he was. Anyhow, if he wasn't the 11.5 for Kingston, he said he was pretty confident he was the 9.32 for Virginia Water, or the 10 a.m. express for the Isle of Wight, or somewhere in that direction, and we should all know when we got there. We slipped half-a-crown into his hand, and begged him to be the 11.5 for Kingston.

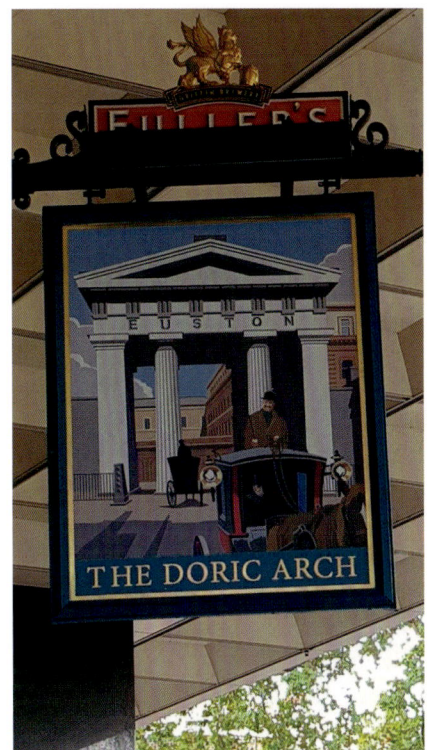

Commemorating the old station, located in the front arcade of the modern replacement, is the Fuller's 'Doric Arch' pub. (*Author*)

Euston's Arch may be gone, but the conceptual grandeur can be seen in the 1823-designed portico of the British Museum, reflecting the Graeco-Roman building rediscoveries of the grand tours of the 1700s. (*Author*)

Right: The feeling of scale of Sir Robert Smirke's design of the British Museum. Euston's arch was criticised by some as being too big and too grand in 1838, for what was then a small station. (*Author*)

Although the Doric Arch underpinned the grand frontage concept, the rest of Euston, while having a certain charm, did not deliver the same grand feeling. You could enter or exit arrivals Platforms 1 to 6 directly from Drummond Street. The left-hand smaller canopies are the original 1838 ones suitably reused and made taller in the earliest LNWR expansion effort. *(Ben Brooksbank CC BY-SA 2.0)*

An atmospheric vista looking south from Drummond Street in 1962. This view along Eversholt (Seymour) Street with St Pancras church in the distance still exists to the left. The shops on the right were swallowed by the 1960s' rebuild of Euston station. *(Ben Brooksbank CC BY-SA 2.0)*

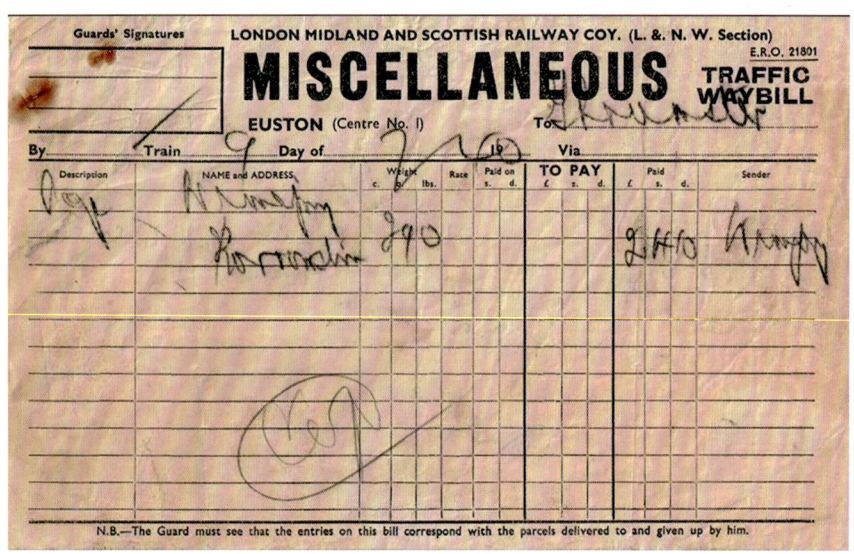

Although the Camden goods station was the central point for LNWR bulk goods, Euston station itself was busy with consignment goods and parcels by day, and newspapers and letters by night. A waybill is a document that accompanies a shipment of goods and contains important information about the contents of the shipment, as well as details about the origin and destination of the goods. The waybill is used as a legal document and serves as proof of shipment and delivery. As can be seen on this LMSR example, it was not always filled neatly. Something for the Clearing House to work out. *(Author's collection)*

While Birmingham, Manchester and Glasgow produced regional newspapers around the WCML, those from London were printed at Fleet Street and distributed nightly from the main stations of London. The stand-alone platforms of Euston's inner station were often used for the loading of vans and coaches for express distribution. When the canopy of the rest of the station was raised by 6ft in 1870 by engineer William Baker, the 'York' platforms bay remained in the old style. (*Author's collection*)

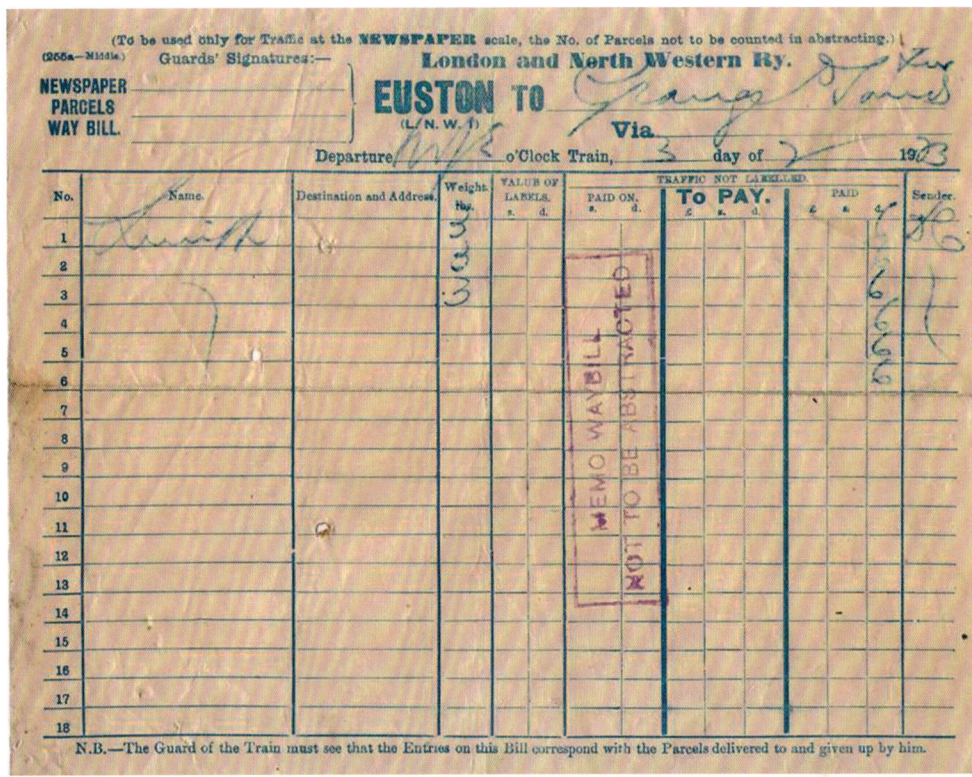

Euston's Platform 9 on 14 April 1961. On the model the shorter Platform 8 accessible from here is as far west as we go. To the right is the Great Hall, accessed where the columns mark the exit from the ticket office. Very much a station within a station, Platforms 9 and 10 mark a small expansion phase for the York Mail service in 1849, prior to building the full western departures side platforms. At night this would have been a hive of activity for newspaper traffic. Platforms on the east side of the station would switch from daytime cab rank to night-time Royal Mail loading facility. This looks more like a provincial terminus than a top rank London station. (*Lens of Sutton Association – KG Carr 4956D courtesy Pfidczuk*)

The LNWR issued its own postage stamps for parcels, as shown here, and had a separate style for newspapers. Postage stamps originated in 1712 as a tax stamp placed on newspapers, which incidentally gave free transit using the postal service. As the tax was less than the normal postage cost, printers developed styles that resembled newspapers to build a thriving industry. In 1855 this tax was abolished resulting in an effective deregulation of newspaper traffic, of which the railways, including the LNWR, were quick to take advantage. (*Author's collection*)

Left: The LNWR made much of the night express in the early 1900s. These trains carried not only parcels and mail, but also had sleeping coaches attached. In the early evening the expresses were set up for dining facilities. These included a named chef for first class at a surcharge. As the night progressed so the carriage formations adapted. At 8pm the Highland Express departed, followed by the 8:20pm Irish Mail bound for Holyhead. The first-class post special (non-passenger) would leave at 8:30pm, then another Scottish Express at 8:50pm. Finally at midnight an express to Manchester, Liverpool, and points north. (*Author's collection*)

Right: Newspaper stamps were specific for the carriage of this printed matter. The London outward distribution of broadsheets was split from the swiftly composited first editions in mid-afternoon to meet commuters in other cities, through to the finals in late evening, to catch the morning breakfast tables of the great and the good. (*Author's collection*)

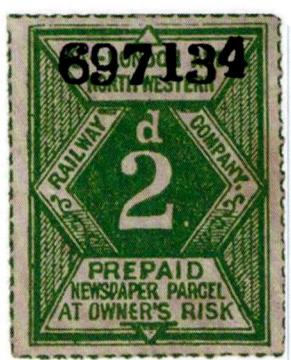

Law and Order

At the foundation of the L&BR there was little detailed signalling as we would know it today. Safety was adequate for the number of services and the speed of trains. Where tunnels or obstructions to line of sight existed there would be a manually operated signal manned by a railway policeman (hence the later nickname of signalmen as 'Bobbies'). The clearance to proceed would be given based on the time interval for the train ahead to clear the area. Such luxuries as block signalling and telegraph communications were to come as speed and technology advanced. Often changes were enforced by government officers following accident investigations.

Signalmen and a true police function became distinctly separate in the 1840s with the LNWR employing warranted officers to detect internal and external ticket and goods fraud, crime involving theft or threats to person, and to protect railway property.

There was illicit money to be made from the railways. Pickpocketing and direct snatch crimes would hit the terminus with organised gangs. The same happens today, especially to the unwary decamping to the Underground stations in rush hour. Euston Underground station in 2016 registered as No. 10 for theft, with King's Cross coming top. The Victorians spearheaded personal insurance for travel covering injuries to person, possessions and luggage. Inbound boat trains were irregular to match the arrival times at port. They were high value targets for thieves to mingle with the confusion of passengers and their possessions.

LNWR police were busy pursuing crimes for false accounting, mis-stating weights of shipments, ticket sharing and reproduction, assault, drunkenness, not forgetting extras such as manning for the Great Exhibition in 1851, Wakes Week travel or football specials. It had dangers. According to the British Transport Police (BTP) roll of honour, ten LNWR police lost their lives on the network, three through acts of violence, fortunately none at Euston. More recently in 1973 the IRA bombed Euston, giving a three-minute warning on 10 September at 1:10pm. The BTP, using loudhailers, were able to evacuate the snack bar area swiftly, limiting casualties to just eight.

On our model we have the LNWR police clearing the Great Hall of passengers to allow the Sultan of Zanzibar to come down the main stairs from a boardroom reception, and then exit to Platform 6 where the red carpet is rolled out.

The census of 1851 listed here is forthcoming in the manning of Euston station on the night of 30 March. The enumerator faced many regional accents, and therefore came up with some interesting phonetic locations. The birthplaces show the early influence of the railway on the movement of the population to London from the North and West. Euston station at this time was approximately half its eventual size. A larger daytime manning would have included a day staff aimed at detective and headquarters functions, as well as security. Later census returns would not disclose the manning details, including one census enumerator bemoaning his lot at having been told in explicit Anglo-Saxon terms to go away.

An example of insurance advertising at the Isle of Wight Steam Railway Museum. (*Author*)

LNWR police officer's name	Age on 30 March 1851	Where born
George Duncan (officer in charge)	34	Parish of Garwood, Scotland [location not found]
Robert Carter	35	Long Itchington, Soham, Warwickshire
William Brooks	28	Yatton, Bristol
Giles Mills	32	Shotton, Flintshire
Samuel Robins	45	Kidlington, Oxfordshire
Jonathan Lorrington	45	Sneinton, Nottinghamshire
Henry Owen	31	St Johns, Camberwell
Jonathan Reeve	24	Staverton, Northamptonshire
William Storkdale	34	Penrith, Cumberland
Thomas Morris	60	Long Clawson, Leicestershire
James Milner	59	Barnsley, Yorkshire

The Station Courtyard

Some areas of a prototype for any model can prove difficult to research. So it was with the earlier canopy. Most images located were of working vehicles or people with glimpses of the structure behind. The alternative was observing the multitude of images of the Doric Arch taken from slightly different angles, and peeping through the columns, as can be seen with this Victorian postcard.

Southern Railway 504M express road horsebox HXW626 at Euston in 1948 as part of comparative post-nationalisation trials. This is a Marathon III variant by Maudslay of Coventry. These were long regarded as the Pullman of transportation for bloodstock. The box made use of a specification like that of the AEC Regal bus chassis and running gear. There was room for two horses via side entry and separate rooms for accompanying grooms. In 1913 the courtyard was infilled at the rear to make a new access ticketing corridor and the steelwork of a new cover expanded into place over the whole central area. (*Lens of Sutton Association 92037*)

Left: Originally Euston had a lighter-weight cover for the courtyard behind the Arch and fronting the Great Hall. We have been able to duplicate this using a mix of 3D-printed items, mainly based on a seaside pier! (*Author*)

The Great Hall

The Great Hall marked a major expansion of station facilities. Opening in 1849, it was executed by Philip Charles Hardwick, the son of the Arch designer, as a waiting room and concourse. In classical style, it was 126ft (38m) long, 61ft (19m) wide and 6ft (20m) high. It incurred a cost of £150,000 (£24.4m as of 2023). At the top of the corners of the hall were eight relief panels by sculptor John Thomas, representing cities served by the line.

The restored ceiling inside the British Museum's Western Hall atrium demonstrates coffered ceiling art at its finest. This leads to the main staircases. (*Author*)

The arms of the LNWR. The Britannia theme continued above the main door pediment to the shareholders' and directors' boardrooms at the rear of the Euston Great Hall. (*Author*)

Above left: The Great Hall in an LNWR colourised postcard of the 1860s. It was based on proportions of a double cub and had fivefold symmetry throughout its length and into the shareholders' meeting room beyond. (*Author's collection*)

Above right: Building the coffered ceiling of the model Great Hall. Foamboard over plywood. To fit the model template the length was reduced by one bay. (*Author*)

The completed cutaway of the Great Hall. Here the police are just clearing the public away from the stairs leading to the shareholders' meeting room doorway. There the directors of the LNWR have entertained the Sultan of Zanzibar, while his Birmingham-bound train is marshalled on Platform 6 (aka the colonnade platform) alongside the red carpet. The original design for the Great Hall was to have been like a Roman bath with semi-circular upper windows. (*Author*)

The real Great Hall as seen on a Sunday in 1962, hence reduced numbers and relaxed passengers. British Railways completed a substantial refurbishment in the 1950s, bringing back some of the glory of this public space. An annual carol concert was held on the balcony and steps, and many exhibitions held either here or in the shareholders' meeting room through the top centre doors upstairs, behind the statue of George Stephenson. (*Ben Brooksbank CC BY-SA 2.0*)

The Ticket Offices and Shareholders' Meeting Room

Above: The light dome for the eastern ticket office is made from half a garden pond fish feeder which nicely replicated the outer glass skin with air vent ports. This could be considered a 'modelling lightbulb moment'. (*Author*)

Left: The upper floor of one of the two ticket offices located either side of the Great Hall. These were contemporary structures within new office blocks. They shared Roman ionic cornice, corbel, and ironwork design detail with the main hall. The eastern side, through which the general express passenger traffic of the line was expected to pass, was the largest, being 60 by 40ft (18.2 x 12.2m). The western was smaller at 56 by 33ft (17 x 10m. It issued tickets for London to York and local branch lines.

A view of the Doric Arch from the north, perhaps only ever seen with any frequency by wartime fire-watchers standing on the directors' boardroom roof. On the right are the clerestory windows for the Great Hall. To the left, the office workers get a wonderful view of water-closet feeder tanks. The twin chimneys come from the boiler room at the base of the faux light well. (*Author*)

Alas, sometimes the best modellers' guide to how a building was constructed is to see it being deconstructed. We made good use of such images to get an idea of spatial associations when making our own decisions. Here is a Watford service on Platform 8 with the site of the Great Hall looking back south to the courtyard, *sans* Doric Arch. (*Ben Brooksbank CC BY-SA 2.0*)

The Rearward Office Extension

To the north of the Great Hall and shareholders' and directors' boardrooms was an open area filled with sidings. Originally this would have been used in conjunction with small turntables to marshal carriages, and overnight for inbound freight to be repositioned. It was ripe for development as the LNWR headquarters' functions grew.

Because of the size limitations of modelling the station and the desire to retain the impressive rear view (one that in real life was only attained again during the 1963 demolition phase), the decision was made to portray a building site. This shows where the first iron columns and girders were going into place based on a discovered plan from the LNWR Society. It covers a baseboard joint where strength was required, necessitating a through bolt, so the sidings are shown here as terminating early as if covered by over-spoil of the builder's activity. It is not quite the correct position for this extension, but it meets a constructional need.

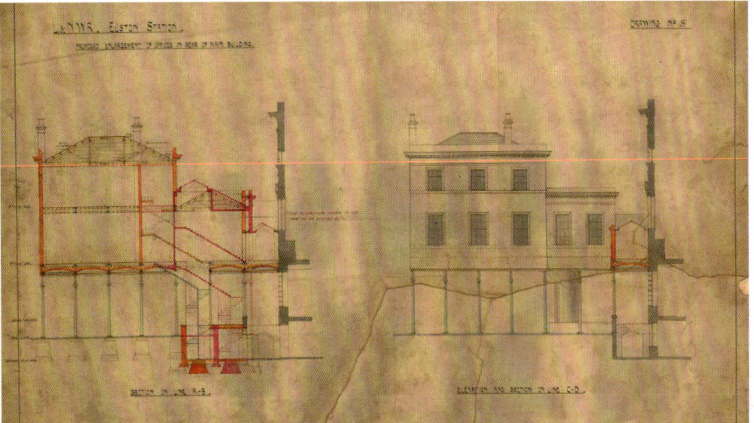

Left: Original 1873 plans for the extension of offices from the rear of the shareholders' boardroom block northwards into the siding space. (*LNWRS DBLDG0252*)

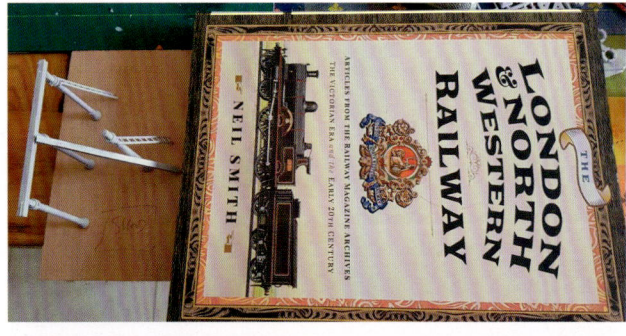

Below: The building site involved raiding the spares box for useful parts. The columns, girders and ladders came from an old Airfix water tower kit. Part of the surface was to represent undisturbed site and utilised coarse grade sandpaper. This was glued with PVA brushed on the underside, then weighted down by a good reference book.

Above and below: Offices were at a premium for the headquarters' function. Operating manager, rules, traffic, customer and shareholder relations, wages, accounting, pensions, strategic planning and commercial were examples of divisions catered for. (*Author's collection*)

Closer to the girders, use was made of old water filter granules (suitably dried out and inert). These were stuck down onto a bed of brushed PVA. Once in place the whole site was painted with acrylics in neutral grey, buff titanium and burnt sienna. The technique is not to mix the colours, just pick up a selection on the brush and swirl it on. That way the chaos of a building site is built up. (*Author*)

Class 3F Jinty 0-6-0T No. 47514 shunting Platform 9 in winter 1959. Behind this, with the limited headroom sign, can be seen the iron columns that supported the office extension. (*Lens of Sutton Association 90865*)

The sidings between Platforms 7 and 8 with the disguised baseboard joint suitably hidden. The grounded coach body used as builder's bothy is deemed to be from a constituent company taken over by the LNWR and scrapped. The flat and open trucks are ratio kits. (*Author*)

Figures are grave diggers by Langley Models repurposed (the associated priest is rather inappropriately in the Euston Square park urinal as a talking point). The contractor's 0-4-2 ST crane tank engine is from the scrap box. You are looking at two Airfix Pug tanks, an Airfix Drewry diesel and some 3D printing mashed up together. (*Author*)

6

The Platforms

The 1951 Festival of Britain special display locomotive No. 70004 *William Shakespeare* arriving at Platform 1, having passed under the Ampthill Bridge, August 1962. (*Online Transport Archive LRTA-GB-BRM-R-4T-11*)

Originally just two wooden-topped platforms or stages existed in the easternmost Euston station, equating to Platforms 3 and 4. Since only three services a day were planned by the L&BR that was more than adequate. Small turntables between tracks allowed for manual and horse shunting of stock out to the adjacent carriage sidings.

As demand grew so did the station and the general traffic usage of the platforms. At times the longer platforms could contain multiple departure trains, the platform staff needing to divide up access points for passengers to route them to the next train. At the time being modelled, Euston was in a state of flux and being enlarged, plus the track layout being altered. This has allowed us to take liberties with the station throat to shoehorn in as much complexity as possible while only modelling one side of the station.

The maps of the station throat on this page show the move away from small turntables as wheelbases increased. The exception is three that still remained on the triple line between Platforms 3 and 4 near the main building dining rooms, shown on a station plan.

There is an extremely wide Platform 2/3 which became the cab/taxi road. Next is what becomes a very narrow platform serving lines 4 and 5. Platforms 7 and 8 are much shorter, serving local traffic with shorter trains. Later these were to become third-rail electrified services. The station grew to the east with new lines for Platforms 1 and 2 pushing up against Seymour/Eversholt Street, and to the west the 'York' Platforms 9 and 10. This appears to be almost an independent station within a station. Finally the land originally allotted to the GWR and a chunk out of the valley side, taking with it a part of the St James's burial ground, made up to platform 15, thus making a departures side to the station. There was also a new turntable added. Cardington Street was diverted to run along the new retaining wall.

When the LNWR platforms were extended to take longer-bogie stock, and more carriages per train, they were economically constructed of precast concrete standing on legs. This allowed a view completely under the platform to the next running line and is best seen on newsreel footage.

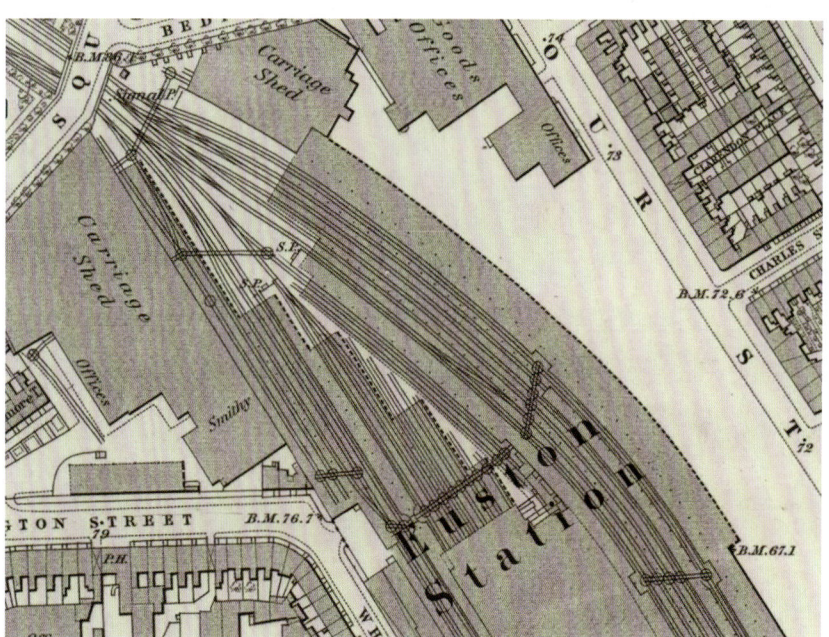

This is the 1876 OS map, accuracy already challenged by fast development of the site. One can only imagine the controlled chaos of running lines combined with alternate traffic of turntable, pinch bar and horse shunting. The local carriage sheds were replaced by buildings further up the Camden Incline at a site now used for the HS2 to emerge into the daylight on final approach.

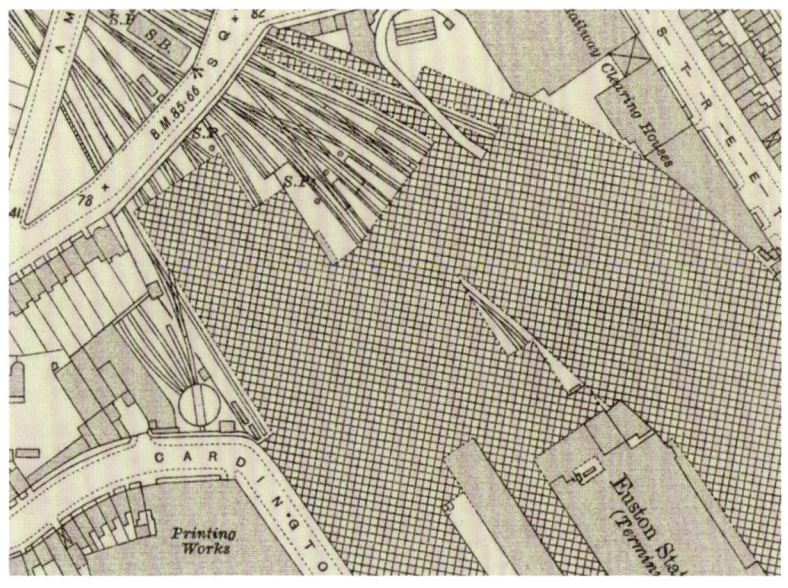

On the 1915 OS map the cartographer has given up and used the glass roof hatching symbol to cover over track and platform details. The site has pushed out the carriage shed buildings either side, occupied with new station facilities. In 1869 the site pushed to Seymour Street for Platforms 1 and 2. An 1883 Act of Parliament was gained to eliminate Whittebury Street and cut the burial ground. Platforms 12 to 15 were opened in 1892.

The down 'Mancunian' sits at Platform 1 on 1 June 1953 having arrived behind 9A Longsight Jubilee Class locomotive No. 45547 *Baroda*. It will soon help propel the empty stock back up the Camden Incline. Platform 2/3 beyond was wide allowing for substantial parcels traffic and boat train arrivals. Down the centre was a taxi road, initially for the horse-drawn hansom cabs and eventually, as seen here, for the Austin FX3 low-loading taxis. (*Online Transport Archive AND-M273-1*)

All change from steam as stopgap to electrification. Sulzer Type 4 Peak Class D8 *Penyghent* is sitting at Platform 1 on 20 April 1961 just after midday. The Bay of Pigs invasion failed on that same day. The Peaks had problems with their steam heating and were rapidly moved to freight as other classes were constructed. This locomotive was named after a fell in Yorkshire and is now preserved at Peak Rail in Matlock, Derbyshire. (*Online Transport Archive AND-M245-2*)

Stanier 8P No. 46228 *Duchess of Rutland* has steam up and is preparing to depart with the 10:25 to Carlisle and Windermere. Originally bridge No. 1 would have stood here, deeply shadowing this view and crowding the station throat. This arm of Ampthill Square was amputated in 1952 rather than pay for bridge deck renewal. It would have been removed for modernisation in common with Ampthill Square bridge No. 2 beyond as it was too low for electrification clearances. (*Ben Brooksbank/'Old Euston': outward on Departure side/CC-BY-SA 2.0*)

The departure-side platforms had a much newer roof arrangement with very little finesse, especially when viewed from the platform end. Here on 20 April 1961, the first of the LMSR streamliners, No. 46220 *Coronation*, blows off abruptly at the head of a Perth service. Alongside, an unidentified English Electric Type 4 (Class 40) diesel readies to pull the midday Scot express with aplomb, but to the spotters present, no theatricality. (*Online Transport Archive AND-M452-2*)

Here is the ticketing circulation corridor at the rear of the courtyard, seen in 1962. This efficiently joined both sides of the station. The Great Hall is to the left, taxi drop-offs to the right. (*Ben Brooksbank CC BY-SA 2.0*)

There were several London Midland Region enamel banners on service chalkboards in the ticketing corridor that was built in front of the Great Hall. Alas, as with so many collectables, the true provenance of this example, such as a site receipt, is long lost. (*Author's collection*)

Automating Model Train Operations

Initially running out to a fiddle yard was envisaged, but the overall length of the layout would be too large for the Club premises. To preserve the feeling of movement while keeping manning levels minimal, the idea of using several shuttle automation modules was proposed. In this manner a locomotive's movements could be limited to within a single 4ft board as a shunting operation. Because this would remove the traditional model layout hinderance of electrical joints between boards, it would make erection of a very large scenic layout faster and more dependable. There are several options available on the market for DC power. MERG (Model Electronics Research Group), Gaugemaster and BlockSignalling were among those contemplated. For this requirement the open circuit board module from Black Cat was decided on. (Note: always handle with care, we dropped one onto a hard floor and lost a soldered chip connection.)

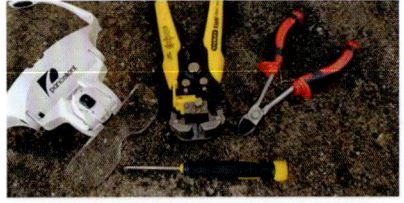 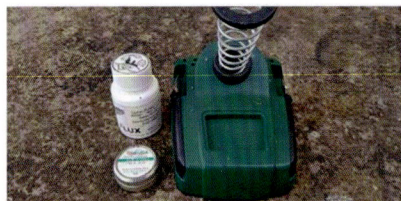 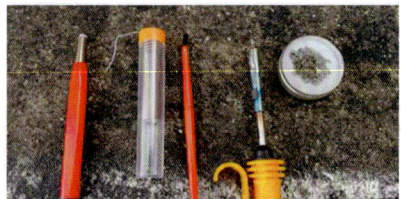

The essential tools for making track wiring less of a task. Left to right: magnifying and lighting headset; wire strippers; wire cutter; small screwdriver; soldering iron safe docking station; tinning solder; weak liquid acid flux; fibreglass abrasive pen; fine solder wire; scrap brush for flux; modeller's fine tipped soldering iron.

Above left, above right and right: Some basic soldering is required to connect the wires to the track and to the distribution board of the module. Ends of multi-core 12V wires are tinned by adding flux then allowing hot solder to melt by introduction of the soldering iron with solder on the end. The track is held side on and cleaned with fibreglass pen, then fluxed, tinned/soldered using gravity. Finally, wire is introduced with more flux. It burrows into the rail-side solder with a satisfying squeaking sound. Wash track after to remove flux residue. (*Author*)

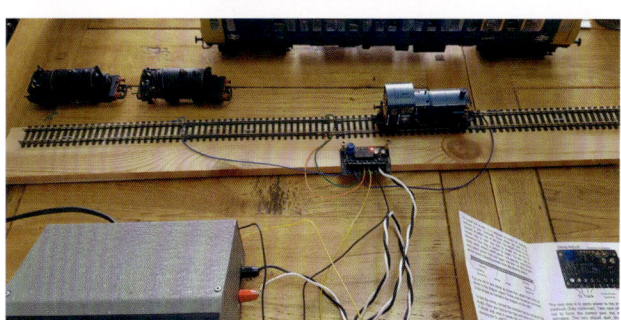

Above left, above right and right: Here on a test jig, we have the wiring in place to demonstrate (colours are just for demo purposes). Yellow and brown wires from the variable 12V feed of controller. Black and white from the 16V accessory feed at the side of transformer. Orange is the earth common return and green is central power. Blue is left isolation section power and purple right isolation section power. When all are switched on, the shuttle will propel the locomotive from an isolation section to the one on the opposite side. Then wait for a pre-set amount of time and auto return it by the module reversing the current. Speed of movement only is set on the transformer dial. We use spare older H&M transformers for each unit. (*Author*)

Above: The winds of change started far earlier than many people suspect. The diesel twins of Nos. 10000 and 10001 were built at Derby just prior to the 1947 nationalisation. The concept was that these 1,600hp diesel locomotives could be worked in tandem on the high profile expresses or singly pulling semi-fast suburban services. In April 1955 No. 10000 shows off the smart black and chromatic silver livery that was applied to both. It is roundly ignored by the train spotters who concentrate on No. 46206 *Princess Marie Louise* making a commotion in the background. This innovator diesel eventually succumbed to the cutting torch at Cashmore's of Great Bridge in January 1968. (*Online Transport Archive AND-M245-2*)

One of the joys of being in a model railway club is the ability to reproduce such an iconic photograph using members' rolling stock. Here utilising a Mainline un-rebuilt Scot No. 46137 *Prince of Wales Volunteers South Lancashire* and the Bachmann version of No. 10001 in early BR livery. (*Author*)

Seen here on Platform 7 in 1959 is Bulleid designed Class D16/2 No. 10201 on a local train. The locomotive was originally over-specified with a 110mph top speed and was transferred from the Southern Region over to Camden shed allocation in April 1955. All three class members, since they were non-standard entities, were withdrawn in late 1963 as electrification reached Euston. They then went on to join the scrapping line at Derby Litchurch Lane Works. They succumbed at Cashmore's of Great Bridge in 1968. (*Lens of Sutton Association LOSA 91009*)

On the central electrified lines in 1959 is an ex-LNWR Oerlikon service exiting the station bound for Watford. They were shedded at Croxley Green and powered from the LNWR power station at Stonebridge Park. These three-car units were withdrawn the next year following the introduction of the replacement Class 501 electric multiple units. In the evening peak service, there were five electric departures from Euston each hour heading to Watford. (*Lens of Sutton Association LOSA 90841*)

Making Coaches

Euston station not only has numerous platform faces, but in Victorian rush hours had a reputation for having multiple trains stacked on each platform ready for outbound passengers. Four- and six-wheel carriages are shorter than their later bogie counterparts. While there were more of them used per train, the overall capacity of the platforms was considerable. Trying to find your correct train in the twentieth-century original Euston was hard enough, even though that had an electronic public address system. With chalkboards, megaphones and barrier-divided platforms, the bustling station in the 1870s would have been a challenge to the hardiest traveller.

As a club this presented a conundrum – we wished to make use of the beautifully liveried Hornby and Hattons ready-to-run LNWR liveried creations. However the overall price involved to fully populate the station was prohibitive. So, a decision was made to have foreground stock in detail and background stock as a conversion.

The regular modeller could consider for example having camping coaches on a station. There a manual repaint of an older style coach would not look out of place, and deeper into the model any irregularities would only show for a well-aimed camera, not the naked eye.

The Hornby children's set 'Bug Box' four-wheel coaches were chosen. The wheelbase and compartment composition represent the earliest carriages. They are relatively cheap at exhibitions, and it is easy to perform a cut-and-shut operation with a fine-toothed modellers' saw. The requirement was to economically create longer four-wheeler and six-wheeler coaches for the model, since the LNWR did not adopt bogie coaches until 1893.

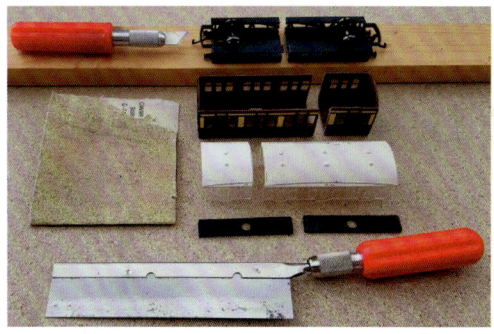

With a four-wheel conversion, two donors have ends removed and chassis cut. Reversing the roof round enables the coach to be stronger as not all the joints are in the same place. For a six-wheeler the headstock/buffers of one donor are cut away and joined to half of another chassis. A whole body is joined to two compartments of another donor. (*Author*)

Examples of the donor vehicles. They varied in livery, paintwork quality and general cleanliness. At times a good scrub was needed before the more delicate modellers' style of preparation. (*Author*)

Chassis, body and an integral glazing and roof were separated out. Rain-strip mouldings on the roof were filed off and the decision made as to whether a four-wheel (four compartment) or six-wheel (five compartment) carriage was to be made. (*Author*)

So far as paintwork was concerned, a fine undercoat of mid-grey acted as the stabilisation layer. The 'plum and spilt milk' livery of the LNWR coaches from Precision Paints was painted into the panel mouldings with a OO sable brush. No attempt was made at lining, numbering, or heraldic crest as these are background makeweights. This example had no partner so simply became a passenger-liveried milk van. (*Author*)

Looking rather like a scrap line photograph of the old GWR broad gauge. Here are eleven examples constructed from sixteen donor vehicles. Fine plastic bridge filler is used here. A gel superglue and granular plastic from Deluxe Materials act to hide bodyline gaps where the saw cutting/filing is not quite perfect. (*Author*)

Above: Non-passenger ready-to-run LNWR rolling stock is thin on the ground. A very nice iron mink explosives wagon exists, which could perhaps be justified for supplying the Honourable Artillery Company with black power for cannon. But the warning label restrictions to consignments probably precluded this traffic to the station. (*Author*)

Above left: As famously quoted, the camera never lies! The underframes are inaccurate, but this detail gets generally lost in the lighting on the layout. Paintwork is regarded as adequate, but definitely not up to close observation standards. (*Author*)

Left: Commercially produced carriages on the Euston model. The Hattons Genesis LNWR range (upper) and the Hornby LNWR liveried (lower). Both are generic but carry the livery well. (*Author*)

LNWR first-class corridor brake constructed 1906 to diagram 127 No. 303 at Wolverton demonstrates the livery as applied to the Royal Train in LMSR days. Seen here on display at Shildon Museum. (*Author*)

Below: A busy station has many small incidents, and most yards would have booked such. Perhaps this should have been placed on all buffers at the station as related below. (*Author's collection*)

Above: A test setup prior to ballasting the station tracks at the rear of the shareholders' common room. This area of track was used for storing carriages, vans and pilot locomotives. The platform at rear is No. 8 and will have fresher ballast reflecting the more recent construction. The station staff nicknamed this area 'Tom Tiddler's Ground' after a short story by Charles Dickens. Commercially produced high detail carriages are in the foreground. The lesser detailed model carriage inconsistencies melt away in the background. Part of the challenge for both a club and for the individual modeller is how deep to go with prototypical requirements. Is an artistic feel and overall portrayal enough? On a model the size of Euston the public quickly become visually overwhelmed, but the builder is always aware of shortfalls and deviations. (*Author*)

Platform Incidents

Over the years the railways tightened up on procedures, technology and rules under direction from the Board of Trade, and latterly the Ministry of Transport, following investigations into incidents. The following are examples of accidents within the confines of the station itself.

On 3 March 1852, a notifiable buffer stop collision took place. This was a result of the main brake wheel being disconnected from the wheel blocks due to rough usage at the head of the grade at Camden. Although multiple guards often rode the train, in fair weather only one was expected to apply the brake.

On 21 February 1888, the buffer blocks of the old station were again the point of impact, this time from a Perth express arriving in platform and hitting at approximately 6mph. Four passengers and a guard were hurt, the buffer plank of the locomotive broken and the buffers of three fish vans, two guards' vans and a saloon carriage bent. Cause was put as vacuum application not made and steam brake late application mid-platform on a late running train.

It was the habit up until diesel and electric days for the locomotive at the buffer end of a train to be uncoupled and assist the exiting coaches up the Camden Incline. On 1 February 1889 the 5:30pm down express, Liverpool-bound, consisting of nine vehicles, was unexpectedly held at the advanced starting signal at danger. Tank engine No. 1919 (confusingly this number later reused on Jubilee Class *Resolution*) ran into the back of the consist. Cause: accidental signal shown against this platform and inattention of both drivers.

On 26 September 1923, the arriving ex-Holyhead Irish Mail hit the hydraulic buffers on Platform 1, resulting in seven minor injuries. The buffer stop split its crossbeam and the bogie wheels of the first vehicle mounted the engine buffers. Coach No. 4 also left the track. Fault was irregular double heading habits. Camden to tighten crew rules.

Poor Platform 1 buffers again proved their worth with the 12:13am Irish Mail overrunning on 29 February 1928. At the head was 4-6-0 Claughton No. 5995. This resulted in eighteen injuries to Post Office workers and passengers. Driver fault due to misjudgement in the curve.

On 27 August 1928, Platform 2 hydraulic buffers had a heavy collision with seventy-four injuries reported with the arrival of the 'Mancunian', pulled by reboilered Claughton 4-6-0 No. 6017. The stops themselves were carried away and the locomotive continued 9ft into the platform itself, terminating just short of a waiting room. Entry speed to the platform was estimated to be 25mph. Listed as driver misjudgement and guard inattention.

On 26 September 1923 at 7:59am Precursor 4-4-0 No. 5284 and Claughton 4-6-0 No. 5980 with a late train overran Platform 1. Recommendation

14 April 1961. Tucked away to the east against Cardington Street are the 70ft turntable, water tower and sand dryer. Note the grounded boiler, possibly from a Renewed Precedent class. (*Lens of Sutton Association KG Carr*)

for the LMSR to look at increasing the buffer retardation in light of so many overruns. The signalmen in No. 2 box, although supposed to spot fast-running inbound trains, were too busy to be depended upon.

On 6 August 1949, an empty stock working comprising that required for the 8:30am Liverpool express was accidentally routed into occupied Platform 13 rather than No. 12 in the arrivals station. The resultant collision was at approximately 5mph, but the 8:37am Manchester train was already loading with passengers. It resulted in thirty-four hospital injuries and ten lightly injured. This was classed as the fault of both signalling and shunter. Overdue signalling improvements resulted in Box 2 being without track circuit indicators for occupation of platforms.

Euston and the Blitz

Following the Second World War the surroundings of Euston showed the scars of war – open bomb sites, patch repairs and generally worn out infrastructure due to lack of investment. This allowed central government reinvestment to include electrification as the eventual replacement for steam and diesel on the WCML. In turn this brought to fruition the replacement of the original station. In this section we cover some of the details of the impact of the Blitz on the LMSR in Euston itself, and in the immediate surroundings.

On the evening of Saturday, 7 September 1940, the Blitz came to London, heralding a period of extensive destructive and disruptive attacks. Initially there were daytime raids that gave the opportunity of targeted bombing on railway stations and goods yards as strategic targets. The advent of the night Blitz following extensive Luftwaffe bomber fleet losses brought in terror bombing of a less discriminative nature (which included the collateral loss of house and childhood Hornby model railway of our club president Colin Brown). During this time hits on railway targets became more luck than judgement. Euston being located north-west of the City of London meant that raids penetrating this far came through anti-aircraft and balloon barrage counter measures.

When reviewing direct damage to Euston in 1940 alone, we also include some of the contributing disruption events to goods and passenger service. It needs to be remembered that line blockages and destruction of sensitive infrastructure – such as stations, bridges, tunnels and viaducts – as headlines also brought other problematic side effects: unexploded and delayed action bombs (UXB/DABs); damaged signalling and telephone lines; loss of paperwork; dispersal of office functions; evacuation; and death and injury of personnel. At times it was remarkable that London got working again, like a kicked-over anthill. To give an idea of the pressure experienced, the total number of devices dropped from 7 October 1940 to 6 June 1941 in Camden alone were 813 high explosive bombs and 16 parachute mines.

The railway executives had advanced recovery plans from the build up to war. Stockpiles of rail, sleepers, aggregates and steelwork were available. They had to be rushed to a location once the all-clear sounded. Temporary fixes were made, and longer-term repairs were planned and prioritised according to severity of damage, and how strategic the rail function was.

The type of air raid could also bring different results. High explosives (HEs) could penetrate and destroy brickwork with blast. Land mines dropped by parachute would explode, blowing off roof tiles, which allowed incendiaries to penetrate buildings. The later vengeance weapons of V1 and V2 rockets, while indiscriminate, could with ill luck hit a critical area. The first major raid had Heinkel He 111 and Junkers Ju 88 eschewing the lighter 50kg bombs and dropping 250kg HEs, 30 per cent of which were delayed action. Set at two to four, and twelve to fourteen hours delay a raid was causing three bands of disruption; 20 per cent of the loads were incendiaries.

On 7 September, the first major air raid didn't quite reach Euston – the East End and docks were the major target. St Pancras was blocked by HEs at Islip Street. On 12 September, the tunnel at Primrose Hill was closed by a UXB blocking Euston totally. It was on Wednesday, 18 September that Euston lost its lucky streak. In Euston Road the front wall of the Somers Town goods yard was blown out (where the British Library entrance is today) and bombs damaged the goods shed roofs. The LMSR Euston House HQ built in 1934 was hit in a later evening raid. This was the same night that the John Lewis department store was burned out in Oxford Street.

On Sunday, 22 September, all traffic stopped due to incendiaries setting extensive fires in Camden goods station. The stables in Prince's Road were also alight but all horses saved. UXBs contributed to delays long after.

During Friday, 27 September in one of three major raids on London, the cutting of Camden Bank received a direct hit between the 'up' fast and slow lines by HE bomb. All traffic was run via the other lines until repaired.

On Sunday, 29 September Camden Town was hit, this time by four HE bombs on what was otherwise a generally quiet raid day. This severed all electric lines resulting in suspension of commuter services from Watford the next day.

A ramping up of damage occurred on 4 October with 270 aircraft on a wide London raid. An HE bomb knocked out Euston No. 2 signal box by Stanhope Street. Signalling damage was extensive, eliminating all traffic to Platforms 7 to 11 plus carriage sidings.

On Wednesday, 9 October, of a 170 aircraft, some penetrated as far as Somers Town goods yard, with bombs damaging both levels of the building and St Pancras Junction being effectively destroyed. Traffic was diverted to Euston.

At a certain point the authorities and the available resources can get completely overwhelmed – 15 October represents this situation, Air Raid Precautions (ARP), National Fire Service (NFS), hospitals, heavy rescue and the railway's own teams were all fully engaged, only for an effective knockout blow to come four days later. Lack of large-scale enemy follow-up on London allowed a protracted recovery of infrastructure and people. Coventry and other cities had taken a critical blow instead.

On 15 October, raids knocked out the Edmonton water channel at New River Bridge. This removed 46 million gallons of water a day from London and the LMSR and other company steam sheds dried up for a time, as did the firefighting effort. The old

Bomb damage map complied in 1945 by the London County Council shows from light to dark the degree of damage sustained by private building stock from the war years. (*UNESCO UK Memory of the World Register*)

channel was hurriedly restored and reversed half of the loss next day. Fast lines were blocked by a UXB at Camden town. Soon afterwards nearly 200 civilian casualties were incurred because of building block collapses at Prospect Terrace, St Pancras and Cumberland Terrace in Seymour Street. Queen's Park Station meantime was heavily hit. This key point for Euston and Broad Street traffic had a triple hit. It lost the roof to the substation, walls came down, blocking all electric lines with 25 tonnes of debris, which in turn was run into by Royal Scot No. 6122 *Royal Ulster Rifleman*, thus derailing the 19:30 Inverness express. This was followed by a burst water main, flooding the line and craters. Full services were delayed until 21 October.

A dark day for the station came on 19 October. By this time, there are contradictory lists of what happened which day and how heavy the raids were. The ARP organisation was severely disrupted, the recording system was playing catch-up while under duress. In the week prior, the author's father, with a heavy 3.7in anti-aircraft gun and searchlight unit was rushed by rail from training in Watchet, Somerset (Donniford Camp) into Hyde Park and latterly recounted tales of hot shrapnel falling like rain during those October raids.

At 20:35 the roof of the Great Hall was penetrated by incendiaries. The hollow construction allowed for walkways through the roof space above the wooden coffered ceiling. Railway employee fire watchers were swiftly able to extinguish the fire. On the western arrivals side and up the Drummond Street, frontage office roofs were set alight. A 250kg HE bomb exploded between Platforms 2 and 3, derailing the 'down' postal with injuries and another severely damaged the frontage of the Euston Hotel west wing. Eventually Platforms 1 through 6 were put out of action, forcing limited use of 7 to 15 until 27 November.

On 18 November the first 1800kg bomb was dropped on London – 358 bombers dropped 414 tonnes of HE and 41,112 incendiaries. Oil bombs and aerial mines were also dropped. The bomber stream target point was Hyde Park. A weather front with intermittent cloud led to bomb creep to the north-west. The Euston arrivals platforms were badly damaged with 14 and 15 requiring extensive repairs. Retaining walls collapsed near the turntable. Also damaged this day were Westminster Abbey, Wellington Barracks and four city hospitals.

Euston itself appears not to have been directly hit in the 1940 Christmas firestorm which wiped out areas from St Paul's to the Barbican.

Records from 1941 show the effects of blast damage and 'down' line blockages but nothing direct. The mini-Blitz of early 1944 was kind to the station. Likewise, the V1 and V2 weapons in 1944/1945 missed the complex although Camden and Somers Town civilian targets proved vulnerable to the randomness of these missiles.

It has to be remembered that London as a whole, as well as the country in general, was under extreme pressure. The LMSR had recorded more than 3,100 incidents with 725 occasions of physical blockage to lines. Unfortunately, 51 LMSR servants died along with 17 passengers; 567 staff and 138 travellers were injured.

The war opened up much of the frontage of Euston for eventual rebuilding without the need for expensive compulsory purchase. The destruction of this housing stock involved part of the Regency villas in Euston Square, and Drummond Crescent behind the Euston hotel, resulting in several bomb-site car parks in the 1950s.

LONDON MIDLAND AND SCOTTISH RAILWAY COMPANY.
Engineer's Department
Return showing authorised Relaying, etc., carried out.
Week ending Sunday

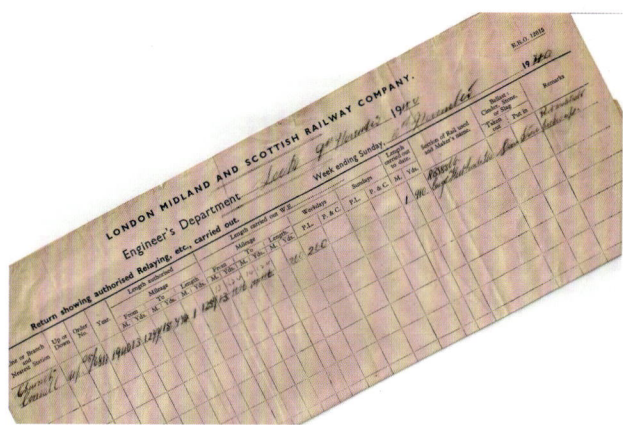

Some maintenance was essential and where possible was carried out before line speed/weight restrictions had to be imposed. Here Leek engineers' department have replaced 910 yards of track in a weekend line possession near Consett in November 1940. (*Author's collection*)

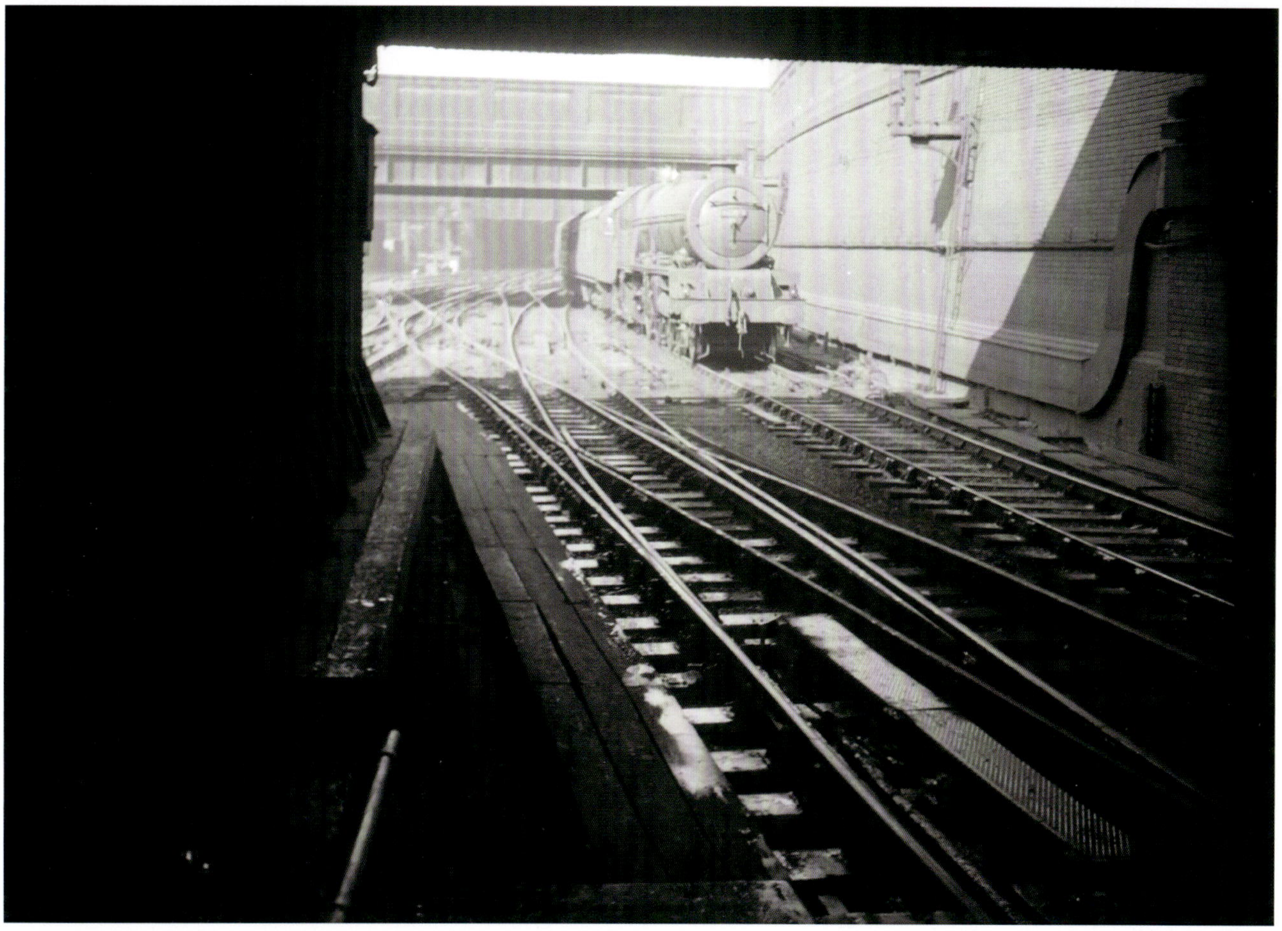

A view taken north from the end of the long thin arrivals Platform 4/5 which terminated just before the Ampthill Bridge No. 1. Here we see BR (LMR) 4-6-2 No. 46204 *Princess Louise* coasting down the bank into Euston on 8 March 1949. This is a far as the Club model of Euston goes, using the Ampthill bridge No. 1 as the eventual terminator to fiddle yard. It was a near identical build to bridge No. 2, seen beyond. The signal boxes for the approaches are beyond here. When compared to 1840s' drawings of the seemingly wide open and clean Camden cutting, 100 years of development, widening and cramming in as much trackwork as possible has made it feel cluttered and confined. (*Online Transport Archive ADav-M001-12*)

Model Tracklaying

A challenge for this layout was laying an inset trackbed topped with a cork layer. We could not pin track in place due to the thinner plywood base, plus it would be unlikely to survive hammer action. Therefore, a technique used on some older layouts was executed. The track formation was glued in place using PVA, and the track weighted down with heavy iron weights until dry (in this case some Club hammers).

Where the track crossed a framework member, we could put a fixing using a long, thin nail. Where board joins occur the aim was as much accuracy as possible. However, given their weight and odd bolting positions some vagaries exist. Positions were determined where full height wood existed under the platforms. Some joints will thus need a quality review if the layout were to run to a fiddle yard in future.

Because of the use of automatic shuttle units, we ensured that chosen lines have clean connections, allowing for some running on this generally static diorama layout. For these the metal rails at a board end are soldered to a pair of brass screws underneath, attached to the frame. This ensured that a good vertical and horizontal rail face alignment was maintained when the board was bolted into place for display.

Other Club layouts have lighter weight more modular frames. For Euston we inherited a methodology and stuck with it for extensions, making a recognisable consistency for whichever members are involved in exhibition erection and breakdown.

With reference to the later Victorian track plan, a simplification was made to represent a transitory phase as we could not show the full station throat. (*Author*)

The heavy Club hammers represented a good dead weight, especially where flexitrack is used. It allowed the PVA glue to set – a length of time determined by warmth, humidity and patience. (*Author*)

For now, it was also decided that pointwork would not be remotely controlled. The shuttles would run on a set route and the concept of a control panel left to a future project if desired. (*Author*)

Finally, the ballast and any greenery set dressing to represent weed incursion was put in place. We used a 50:50 dilution of PVA glue with washing up liquid added to break surface tension. It was applied using a child's liquid medicine dosing syringe. In theory part of this area had wood or granite setts rather than ballast. (*Author*)

7
The Railway Clearing House

Above: Seen here in March 2023 is the site of the Railway Clearing House (RCH) in Eversholt Street (previously Seymour Street) which is now occupied by Royal Mail. The model of Euston has the original buildings portrayed as they infill a triangle of land as the station throat narrows. No. 163 and successive buildings in that terrace are built to a similar style to the original building the other side of Barnby Street and represent the extension offices of that august body. (*Author*)

Right: The 25in to the mile 1915 map shows the RCH original buildings shaded red, infilling the triangle between Barnaby Street and Seymour Street. The cross marks the goods and later through traffic access gateway. The progression north up Seymour Street can be seen as reflecting the growing importance and complexity of the pre-computer paperwork of the RCH.

The Seymour Street frontage of the RCH on the Euston model seen from the north. The furthest building dates from the 1840s, when the RCH was founded. As the national railways grew so did the workload. Spreading northwards, the next extension was in the later 1850s with an entrance road. This was utilised by Euston station as a vehicle exit once the cab road was built on the wide Platform 2/3. This frontage has extensive windows, required to allow light into the clerical workers' area. Lower frames were rarely opened to prevent disastrous paper-moving draughts. Behind the railings the basement half windows use tin foil to reflect their image, thus, to appear to be complete in a deep light well. (*Author*)

RCH was established in 1842 performing an essential duty as the founding railways of Britain amalgamated or shared running rights over each other's metals. A method was required to share revenue fairly in accordance with agreed costs and universal standards.

The first railways were not joined up – you would travel from London to Birmingham and then purchase another ticket for onward travel to Manchester. As the network and ownership grew in complexity, so did the need for a comprehensive back office solution, and

The 1848 building of the RCH is closest to the camera on this view of Seymour Street looking north. Beyond the carriage entrance and up to Barnaby Street is the first expansion building. As the organisation grew it progressed further up the street. The later buildings are still visible there today on the now renamed Eversholt Street. (*Amalgamated Press*)

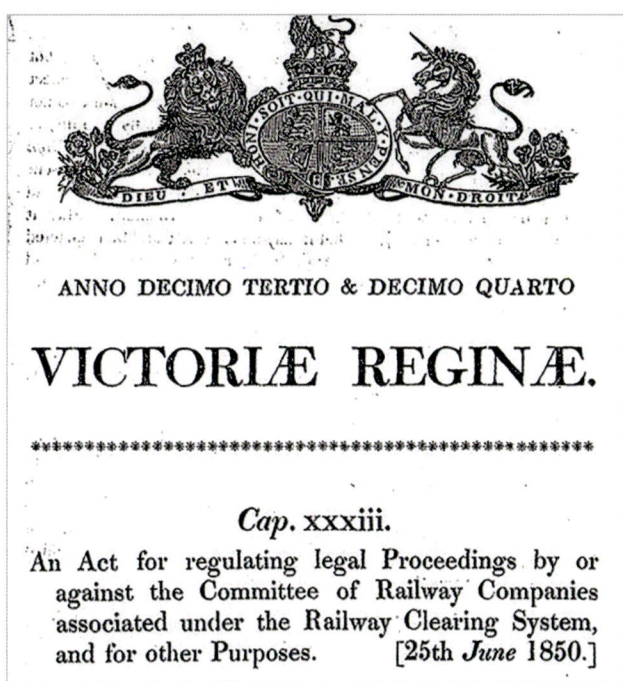

thus towards a seamless railway industry experience for the customer.

Adoption of broad gauge, such as with the GWR, forced transhipment of passengers and goods from one company's carriages and wagons to another. Such rehandling activity was expensive and could be avoided on the routes of the standard gauge companies, once agreed rules and renumeration had been established.

It is thought that the L&BR board was instrumental in the establishment of the RCH. Chairman Lord Wolverton (Mr George Carr Glyn who also was a banker and familiar with the Bankers' Clearing House), engineer Robert Stephenson (son of George Stephenson) and Kenneth Morison, Chief of Audit, were all from the L&B. Upon foundation on 2 January 1842, they employed just four persons and were ensconced next to the Victoria Hotel on Drummond Street opposite the Euston Arch. The potential of the organisation was soon seen nationwide and £193,246 in receipts were handled in that first year. By 1848 the RCH needed to expand into a purpose-built facility tucked into the top north-east corner of the Euston station site. This and the first phase expansion buildings of the early 1870s are modelled within the Euston diorama.

Initially, nine railway companies participated in the organisation. These were the L&BR, Midland Counties, Birmingham and Derby Junction, North Midland, Hull and Selby, Manchester and Leeds, Leeds and Selby, York and North Midland and Great North of England.

Just three years later the number of companies utilising the RCH hit sixteen with 656 route miles covered, and consequently the monies handled greatly increased. Not all companies were solvent or had spare money during the age of expansion, so aged debt built up quickly. Eventually they required a private Act of Parliament in 1850 to allow the company to sue its members in court for financial delays and shortfalls.

The RCH earned money from operating company subscriptions and therefore did not take any percentage of the gross. This brought stability to the system as there was a fixed budget that did not vary according to receipt variations per year.

The organisation was broadly divided into three working sections: merchandise, coaching and secretarial.

The merchandise department worked with goods, minerals (at this time including coal)

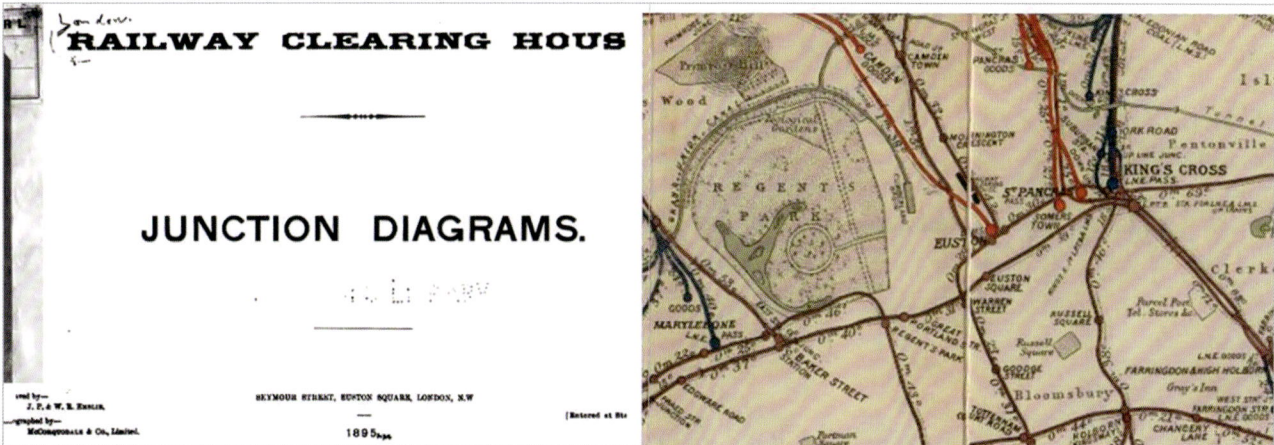

The RCH issued numerous revisions of track layout and ownership to accompany regulations, as shown on this 1895 extract of northern London termini.

and livestock receipts. The dispatching station and the receiving station would both have submitted receipts along with notification of any payments received. Terminal charges of collection and delivery would first be removed. Then the total mileage of the transit over each involved company line would be tabulated. This is the primary reason that the RCH published its famed sectional maps of Britain showing lines, junctions and ownership.

The coaching department was responsible for cross-boundary passenger tickets and excursions. It collated station ticket office sales and traffic returns with ticket stub receipts from destination stations. After a series of netting exercises of credit and debit for each company a settlement was made on a regular basis.

The secretarial department kept the archived accounting materials (muniments) and records of settlements for the use of a company's carriages and wagons on traffic outside of that company's network. Mileage (usage) and demurrage (sitting empty unreturned) charges were raised on wagons and their covering sheets. Wagons at the end of a journey were unloaded and sent back empty to the forwarding company at an agreed junction, and at the same time would receive from that company its own empty wagons.

The following worked example is for a wagon load from Midland Railway (MR) Bristol to London and South Western Railway (L&SWR) Southampton via the Somerset and Dorset Joint Railway (S&DJR), which would invoke Joint Passage Agreement charges, fixed tolls for tonnage and mileage charges. On making a settlement there are three companies to share in the receipt of £10: the MR Bristol to Bath, the S&DJR Bath to Templecombe, and the L&SWR Templecombe to Southampton. All is expressed in pounds, shillings and pence:

Terminal collection/delivery charges on 10 tons at £1 per ton

To Bristol St Philips (MR) at 2s 4d per ton the station-to-station rate £1 3s 4d

From Southampton Docks (L&SWR) at 2s 4d per ton the station-to-station rate £1 3s 4d

Total £2 6s 8d

The remining amount of £7 13s 4d would then be apportioned according to the distance on lines taken, in this case a nice round 100 miles.

MR 10/100 £0 15s 4d

S&DJR 30/100 £2 6s 0d

L&SWR 60/100 £4 12s 0d

Total £7 13s 4d

Resulting in a final balance to be paid to each company as follows:

MR £1 18s 0d

S&DJR (no terminal charges) £2 6s 0d

L&SWR £5 15s 4d

Total £10 0s 0d

Another facet of the RCH secretariat was the establishment of standards for rail vehicles such as tankers and product packaging. In the 1930s nearly 180 meetings a month were held in the extended Seymour Street premises on topics

(Author's collection)

ranging from postal package standards through product crate and box packaging to the agreed lot designs for wagons. Meeting rooms and a high-quality boardroom were in frequent use.

Prior to the Post Office (Parcels) Act 1882 the railways rather than the Royal Mail were primary shippers of small parcels. Until that time the Postmaster General was obliged to negotiate with the RCH for terms of carriage, since the Post Office was only permitted to handle letters and small packages.

The Post Office (Parcels) Act was significant because it enabled individuals and businesses to send and receive larger packages through the mail, making it easier and more convenient to conduct commerce over long distances. This was particularly important for rural communities and small businesses, which previously had limited access to transportation and shipping services. The Act also had a major impact on the Post Office itself, which saw a dramatic increase in parcel volume and revenue in the years following its passage. It helped to establish the Post Office as a key player in the transportation and delivery of goods in the UK and paved the way for the development of modern parcel delivery services.

Agreed standards were required ensuring safety when shipping and handling combustible and explosive materials. At the other end of the scale, they existed to ensure that when an insured breakage in transit occurred it was in a container that met a registered design. If not then an insurer was less likely to pay in full.

The rear of the Seymour Street premises therefore was a temporary store for physical examples to aid discussion of everything from acid carboy bottles with straw packing and wickerwork baskets through to large packing crates suitable to ship the latest machine tool products. In the model itself we have incorporated these, as the taxi ramp exit from the station did not yet exist.

Communications and negotiations were also carried out by the RCH secretariat centrally representing British concerns with individual continental railways to promote international shipping, agreed pricing, service timings and renumeration. The global railway industry had to wait until 3 May 1922 to get to an International Union of Railways (UIC) to act as a coordination body.

The model of the RCH was designed and built using a combination of town planning scale online maps via the National Library of Scotland, plus aerial views prior to 1939, such as the Aerofilms examples included in this publication.

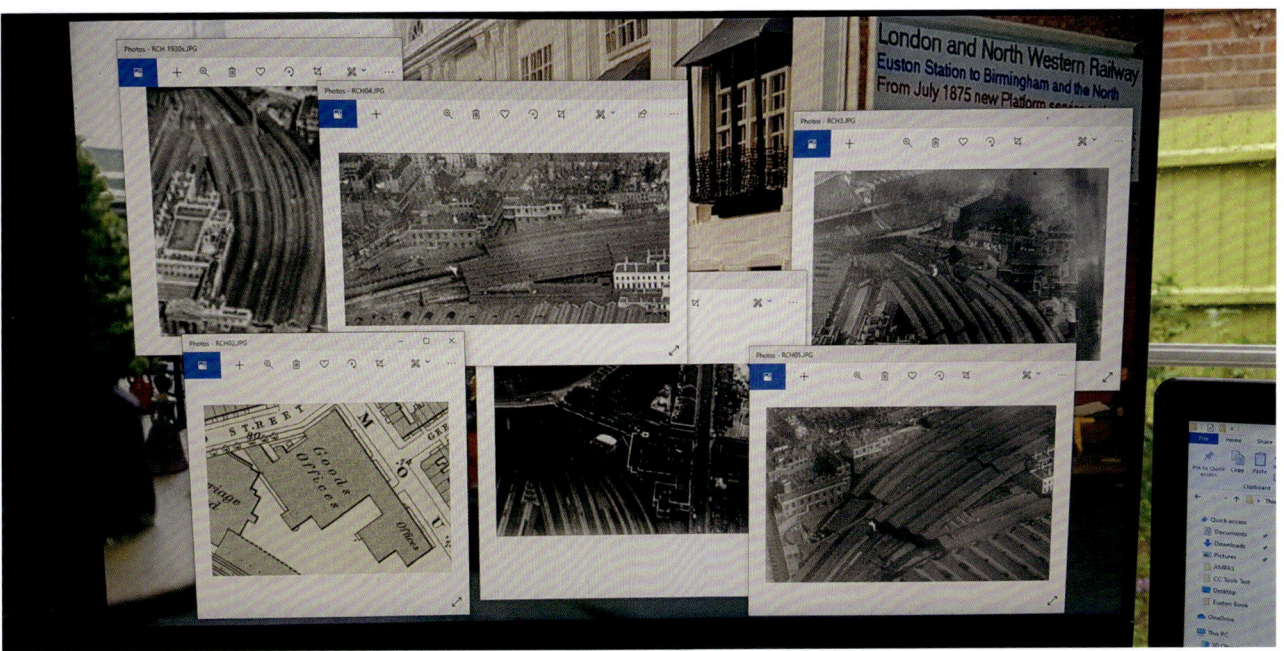

Many online sources were used to get the detail of the RCH Seymour Street premises as it was always overshadowed as an entity by the adjacent Euston station complex. Street and aerial photography, if not reproduced commercially, can be reviewed online, likewise contemporary maps for research. Modern technology has been a boon for modellers, whether for gathering primary evidence, learning new techniques or using 3D printing to originate components for their own requirements. (*Author*)

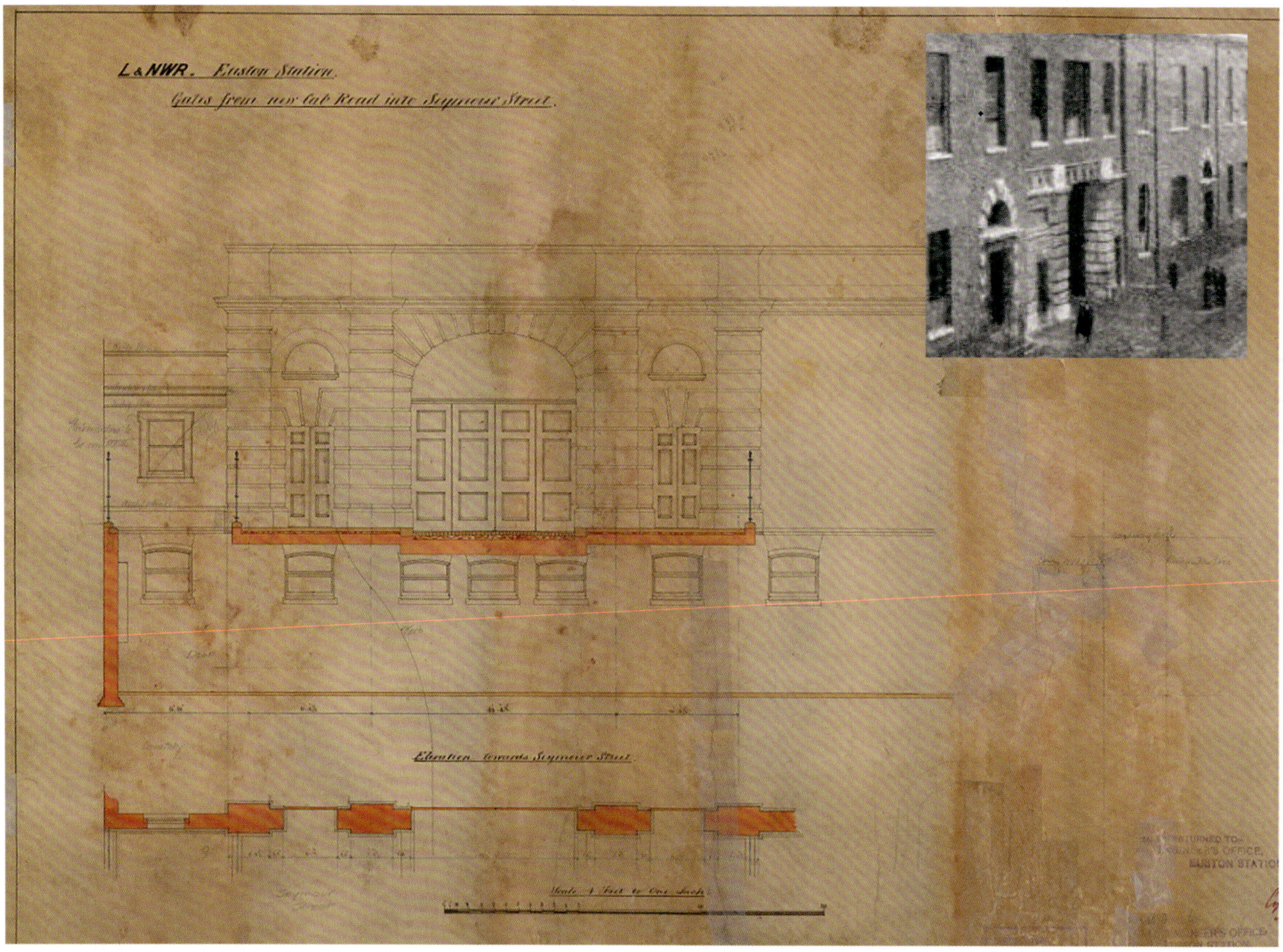

Once the cab road was built, the exit into Eversholt Street required an upgrade in gates to be able to close the station. This original plan shows that some façade updates were planned. The inset image from the 1930s shows that it differs in detail. When building a model of a known prototype/location, your primary evidence may not always be as accurate as it at first appears. (*LNWRS DBLDG0419*)

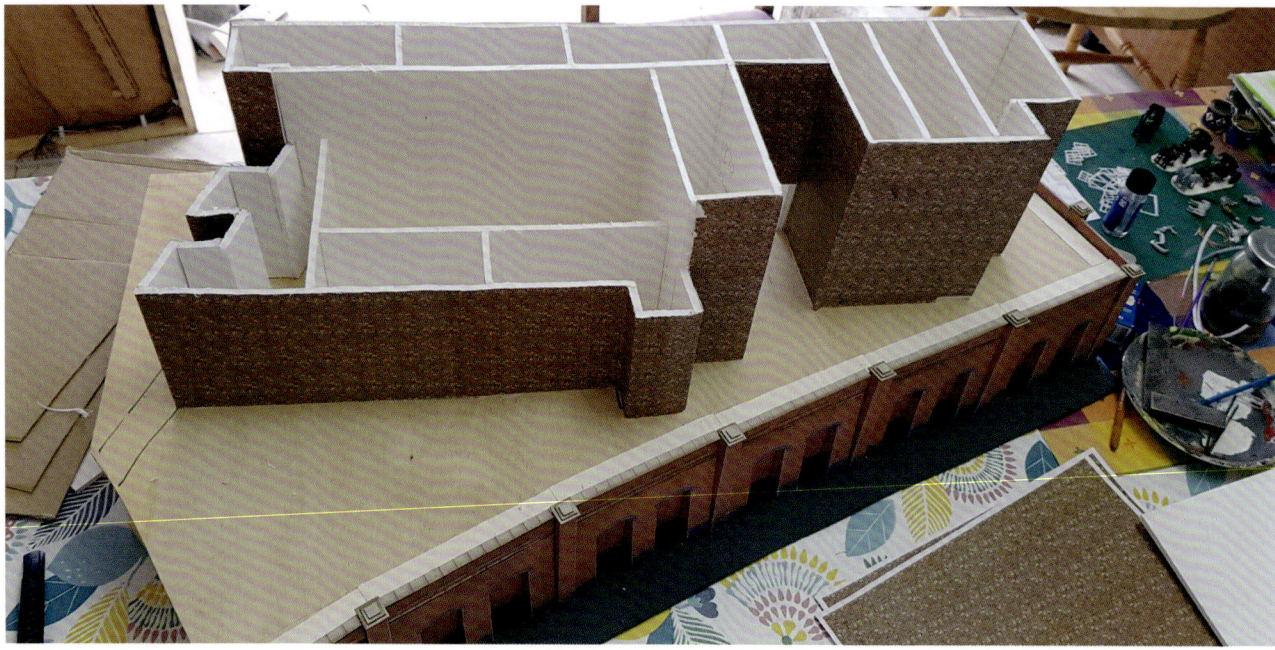

This is the RCH under construction from foamboard with brickpaper covering. This was followed by details from 3D-printed overlays of windows and architraves. Seen here above the northern end of the Euston Platform 1 retaining wall (a Metcalfe card kit). (*Author*)

The cab road exit as modelled with earlier wicket gates. An attempt to add the keystone design on the door surrounds on this elevation felt very crowded and detracted from the model, so the decision was made to deviate from the prototype here. A large model is often a chain of compromises, exact dates to be portrayed, available evidence and the need for truncation and skewing to fit the site. The hope in the end is that at the risk of offending some purists, the build team has survived and is still talking to each other, while as a cohesive whole the model gets the general story across. Think of a photograph versus a painting. (*Author*)

The Barnby Street elevation of the RCH. This is where the visitors would enter for intercompany negotiations and board meetings. The rear yard overlooking Platform 1 of Euston has a 3D-printed grounded coach body of the 1840s stagecoach construction style. The rear yard was used to store barrels, crate packaging and other samples. These were presented by companies for the purpose of submission to gain an RCH 'passed as standard' declaration. (*Author*)

The Stockton and Darlington carriage in the National Railway Museum (NRM) Shildon gave the idea for the RCH grounded carriage body. Composite No. 31 was built in Darlington in 1846. Lightly framed and built on the horse-drawn carriage body principle, these carriages dated quickly as technology advanced. (*Author*)

The RCH secretariat muniments room in the 1930s. Documents pertaining to rail banking financial instruments, legal arrangements for track passage rights, approved diagrams, directors and company emoluments were kept in the records. (*A.P.*)

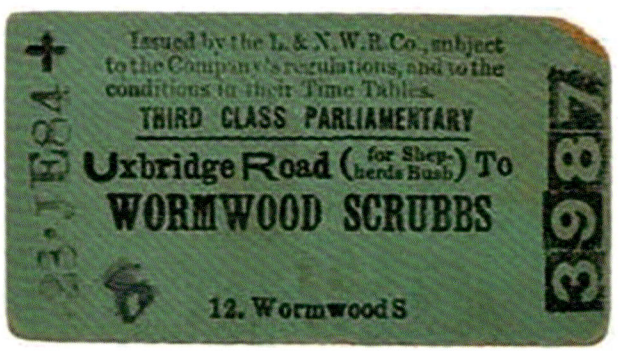

One of the counting offices at Seymour Street. Despite the electric pendant lighting it still has a Victorian feel to it with the chief clerk sat in a reserved corner. The provision of large windows shows that good daylight was required to perform the daily toil of ticket and triplicate carbon counterfoil matching. (*A.P.*)

Counting the stubs of Edmonson tickets collected by inspectors at the exits of stations around the country. The Edmonson board ticket of standard size, colour and typeface with set serial numbers was introduced nationally in 1842 by the RCH. They lasted in service until 1988. (*A.P.*)

Ticket inspectors punch and examples of the steel print plate for producing the Edmonson board tickets. The author has a childhood memory of a cheap day period return discount ticket from Teddington south-west London to Harlech on the Cambrian Coast via Birmingham New Street. By the time everyone in officialdom had zealously punched it there was very little remaining to give the route and fare information. (*Author's collection*)

THE RAILWAY CLEARING HOUSE • 81

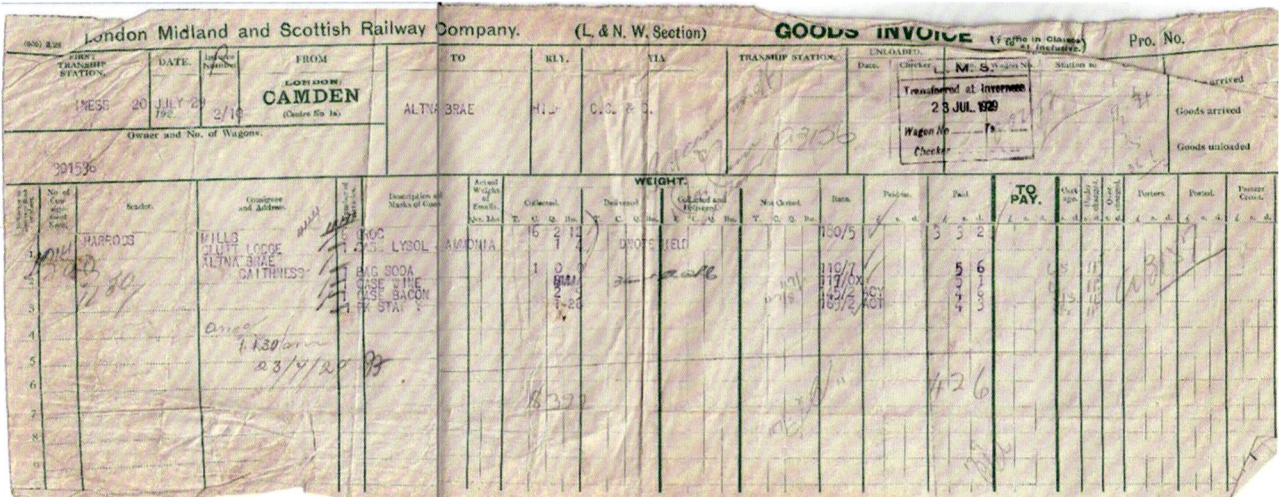

Eventually documents began to contain more useful information and were carbon copied on a typewriter as is demonstrated on this 28 July 1929 Harrods invoice from Camden goods to Altna Brae (Altnabreac Station) in Caithness. Still mainland UK, but almost at Thurso and Wick. A long way to send quality groceries, Lysol and ammonia, soda, wine, bacon and a pack of stationery. Based on recent passenger volumes, this was the UK's twenty-eighth least used station in 2018, yet the LMSR had a goods service running. (*Author's collection*)

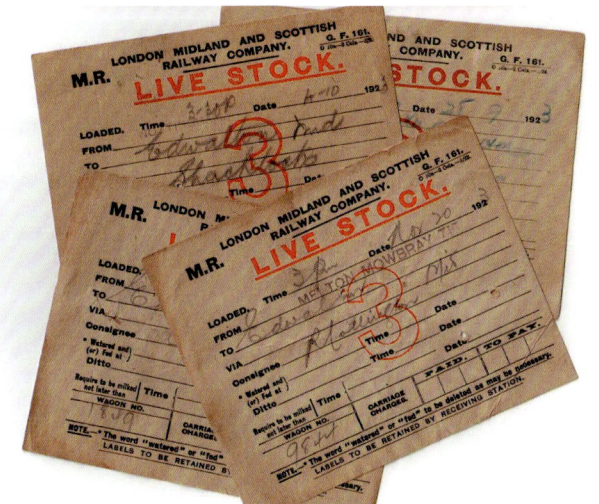

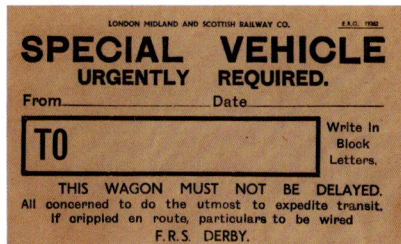

Above: The common carrier rules led to some interesting traffic. Pigeon racing from remote stations was lucrative. (*Author's collection*)

Left: Physical wagon consignment tickets would be applied based on the paperwork generated. One tends to forget that some loads may need to be fed, watered and milked en route … (*Author's collection*)

Left: … while other customers may be of a stature demanding that the best of services were expedited for a specific load. (*Author's collection*)

Below left and below right: If a consignment was not packed to RCH rules it could still be accepted. However, much small print existed to obviate blame from the carrier in the event of spoil or breakage. (*Author's collection*)

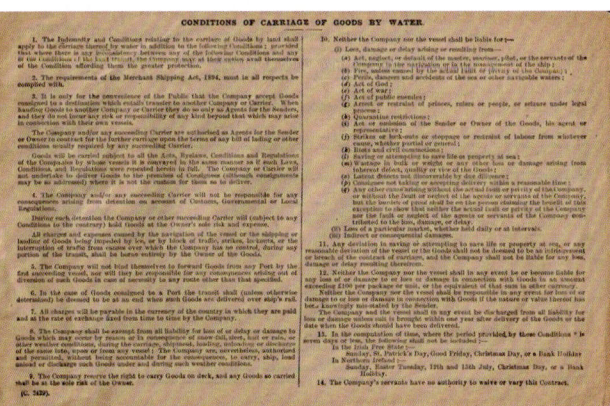

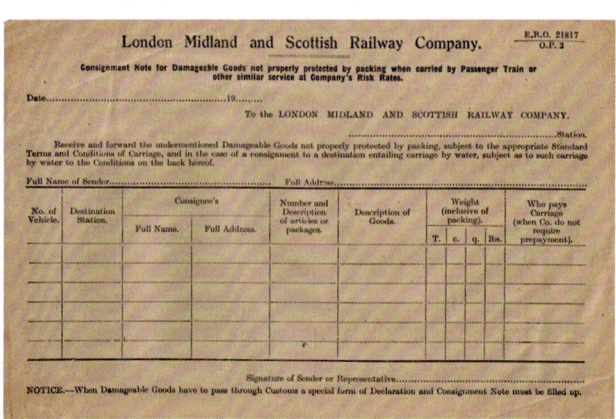

8
Under Euston

Seen on 2 November 2022 on Drummond Street is the blood-red tiling of the Leslie Green designed City and South London Railway (C&SLR) station entrance. Alas this survivor is within the HS2 demolition zone. Built in 1907 it survived closure because of the old lift shafts acting as ventilation for the Tube lines below. The building is effectively hollow with little original interior detail surviving. (*Author*)

The Cut and Cover Underground Railway

The Metropolitan Railway was built along the Euston Road as a 'cut and cover' style of construction, resulting in the station and running tunnels being directly under the throughfare. The first sods were cut in the south-east corner of the eastern Euston Square Gardens to build a service access. The buttressed support walls of the station allowed a 45ft 1in (13.74m) span of bricked arch to be made. In common with Baker Street, that can be seen today, there were white glazed tile covered light and smoke vents, a number of which came out in the gardens of houses fronting the road.

The Gower Street (latterly Euston Square) station was opened on 10 January 1863 with Great Portland Street to the west and King's Cross St Pancras to the east. It has always been disconnected from the main line station. A proposal to construct a connecting subway was cleared by Parliament in 1890 but dropped. As part of the HS2 scheme once again it is on the table as an integration option.

A woodcut illustration of Gower Street Metropolitan Railway station. While natural light would filter through, and gas lamps provided night-time illumination, there was a stygian gloom, especially when a condensing locomotive stood in the station. (*Artist unknown*)

The platforms were originally constructed of wood which did not raise any worries with live steam locomotives. Competition from the electric powered deep tube resulted in a loss of 4 million passenger journeys by 1900. This resulted in the Metropolitan and District companies adopting electric traction fed by a new power station at Neasden. Local and Inner Circle trains were multiple units, and longer distance services using British Westinghouse (and later BTH) manufactured Bo-Bo locomotives hauled traditional bogie stock. All platforms along the lines were rebuilt in concrete as a fire precaution against electrical sparking.

The Deeper Tube Lines

The geology of London lends itself to tube line excavation due to the abundance of soft clay strata. While it presented stability issues, these were overcome by using the Greathead shield to protect and stabilise the cutting face and insertion of precast tunnel sections. One even survives today in disused tunnels below Moorgate station.

Although the ability to dig the tunnels had improved, it had to await the advent of efficient electric traction to replace steam (acceptable in the shallower cut and cover with numerous ventilation points) and displace the interim concepts of pneumatic or cable driven carriages.

The world's first deep electric-driven tube line was the City and South London Railway (C&SLR), which latterly became the Northern Line of London Transport. While this was established in 1890, it took until 1907 for the

One of the capped-off lift shafts now used to ventilate the Tube stations under the Euston complex. Induced pressure ventilation does not exist here. As trains move through the running tunnels, an eddy and flow of hot and cooler air cycles through the old service tunnels from these lift shafts. Where dead-end tunnels exist with electrical equipment present, there is a dramatic temperature rise within a zone of stale air. The cast iron segments of the lining are clearly shown. (*Author*)

Above left: A connecting subway bridge over the C&SLR running lines, closed in 1963 as the main line station was demolished. (*Author*)

Above right: A C&SLR ticket booth is a rare survival in one of the old, closed Euston transit tunnels. In the days before amalgamation by the London Passenger Transport Board, through tickets and fares were sold between different companies. As through travel did not result in a ticket stub for charge back to the originating company, it is thought that a window such as this allowed transit between the different lines, by giving the traveller an onwards travel ticket. (*Author*)

line to reach Euston station. The Walter Scott and Middleton company was contracted to excavate what was then to be a terminus with a 180ft (54.86m) island platform. This had an onwards tunnel leading to a locomotive traverser enabling run around of the traction unit. Early cooperation led to a subway being dug to the Hampstead Underground station and the three lift shafts, one of which also included a spiral staircase to the surface as well.

A new ticket office was opened facing onto Seymour Street to the east of the station, adding to the complexity of the tunnelled workings. There was a new signal box (one of nineteen on the line). This worked on a modified absolute block with such features as gas-lit lenses without signal arms in the restricted 10ft 2in (3.10m) diameter space of running tunnels. With station-based warning treadles activated by trains, these boxes could pass twenty-eight trains in one hour.

Examples of 'time capsule' advertising preserved in the service tunnels. The Blue Pullman and an early example of the Inter City 'arrows of indecision' exist alongside other colourful examples. On the Underground the colour red remains vibrant as it is away from UV light. (*Author*)

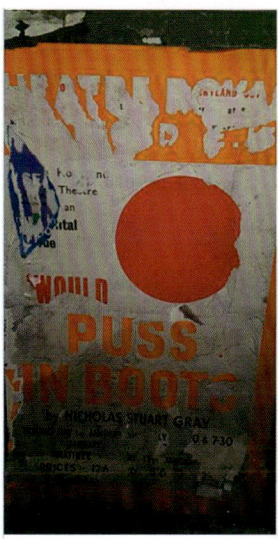

Above left: A rather abstract view of a Victoria Line train in platform as seen from the higher service tunnel ventilation duct. Approved in 1955, the Victoria Line was formally opened 7 March 1969. Over time the London Tube tunnels have heated up the surrounding substrate from passenger body heat and train brake friction. Thus, the Tube has moved from a cave-like ambience to the extreme heat we know today. (*Author*)

Above right: The mainline station of course could be said to have an underground, there being split levels for parcels and goods handling. Here are the Motorail and parcels platform ramps of the western station as seen in 1991. (*Ben Brooksbank CC BY-SA 2.0*)

Aborted Schemes

Euston as a major overground terminus generated much onward travel. Originally the horse-drawn omnibus and hansom cabs served to supplement foot travel. Mass transit schemes were not long in being proposed to Parliament, but many fell to the wayside due to lack of capital, competition or being proposed just as a spoilers to another company. Because such railways would require major infrastructure, non-completion of a scheme was not contemplated. Capital had to be fully raised by a combination of taken-up share issues, mortgages and independent subscriptions (preferential shares) before works could commence.

In 1885 a cut and cover railway was proposed for the area by the Charing Cross and Euston

Railway Company. This would have been a branch from the LNWR at Hampstead Road, curving to pass south under Cardington Road and Drummond Street. It would then curve parallel to the Euston Road and under the Metropolitan Railway but, short sightedly, had no interchange planned. The idea was to rent services to main line companies rather than run their own. The bill was quickly withdrawn.

Around 1891 another north-south railway using the cheaper cut and cover method in the more expensive city area was proposed. The Euston, St Pancras and Charing Cross Railway (subsequently London Central Railway) would have had a branch to Euston following Drummond Street and Chapel Street from the eastern Euston Road. Interestingly, east of Seymour Street the infrastructure plans defined tunnels on top of each other due to the narrowness of the roads heading to St Pancras and King's Cross. Although demolitions and land purchases for stations took place, agreements with the London County Council (LCC) for subsidy failed and the scheme died in the early 1900s

There was a plan submitted to Parliament in 1899 for the Baker Street and Waterloo Railway (BS&WR), proposing creation of a branch to Euston from under Regent's Park. This would exit the park bounds under the Cumberland Gate going below existing tunnels to terminate at Cardington Street to the west of the station. Onward tunnels under Seymour Street would allow stock and traction to be able to swap platforms. Despite a prospectus, shares were never issued, and this scheme quietly withered on the vine.

In 1911 the LNWR decided to divide traffic up and relieve stress on the terminus. It was realised that passengers on the shorter commuter services desired destinations beyond Euston. There was a threefold plan. At Queen's Park there would be the ability to disembark and take a North London Railway train to Broad Street; also the option to take a train to the West End and terminating on the other side of the Thames at Elephant and Castle (later this became the Bakerloo Line). The third option, proposed but abandoned, was for the LNWR to have its own loop line burrowing under Euston and catering for Watford and other services. Several lines serving central platforms at Euston were electrified for taking this service as a cost-effective alternative.

At the start of the Second World War the BBC was looking for dispersal of key functions into strongholds away from Broadcasting House. At that time the Northern Line Drummond Street entrance to Euston Station was disused and was surveyed for the purpose. This plan was supplanted by an abortive scheme to drill two tube-sized tunnels from Oxford Circus towards Broadcasting House. Promoted by the Tube as it assisted the plans to enlarge that station, it was rejected by the Treasury as too expensive. A citadel was eventually built under Broadcasting House in a project similar to the cabinet war rooms.

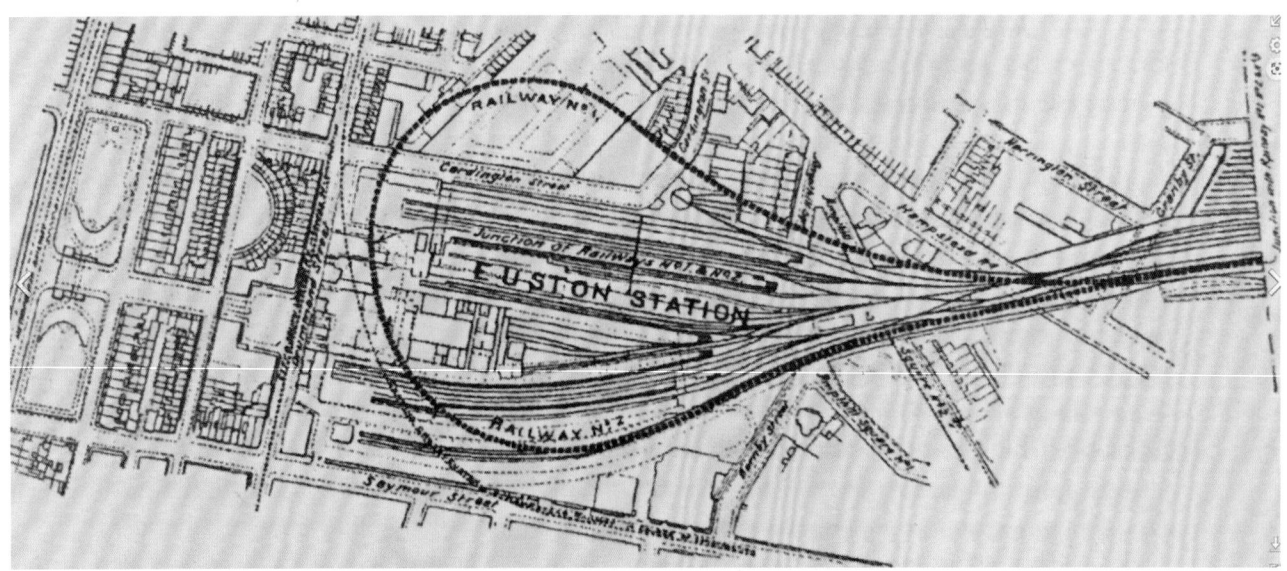

The submitted proposal for the LNWR loop line under the main station. (*LNWR Railway Act 1907, papers public domain*)

9

The Locomotive Sheds

The Main Service Sheds of Camden and Willesden

No coverage of Euston as an active entity would be complete without mention of the sheds of Camden and Willesden. They kept the beating-heart terminus of the Premier Line operational. It is easy to forget their importance in these days of electric traction, where motive power is spread along the bogies of a unit, with a cab at both ends and high availability compared to steam. Locomotives needed fuelling, watering and frequent raking out of ashpans to get rid of clinker, as well as clearing the smokebox of soot and debris. Company prestige demanded thorough cleaning. Intermediate servicing was also more frequent. The steam locomotive was often held up, during the days of the British Transport Commission (BTC) modernisation, as something both robust and dependable when compared with prototype diesels.

In Euston the arrived train would be withdrawn to the carriage sidings for cleaning, freeing up the locomotive at the head to go on shed. When being turned around promptly, about ninety minutes was needed for coaling, watering, a mechanical check and oiling. On the departures side empty coaching stock would be shunted in ready for the top link motive power to couple up.

The biggest regular pressure on the motive power division was the summer holiday season, where numerous extras and charters were run on popular services. The back ends of sheds were trawled by the foreman for

An unknown Camden shed fitter dwarfed by the cylinders of Coronation Class No. 5123 *Princess Alice*, (named after the Duchess of Gloucester), 8 March 1949. This locomotive was originally streamlined and received smoke deflectors in August 1946. Pride in the job shows in the stance. (*Online Transport Archive ADav-M032*)

likely locomotives and crews, matching them to life expired and careworn carriage stock retained for such purposes. Willesden freight locomotives would be called upon in extremis, not needing carriage heating in the summer months.

Irregular stress came in time of war, when the LNWR ran special troop trains for 7.3 million journeys – for example, the runs for the Admiralty organised by Captain Rowley from Euston to Thurso, called the 'Jellicoe Express' aka 'Misery Express', which served the fleet anchorages at Rosyth and Skapa Flow. These trains carried 500,000 men and covered 388,700 miles from 1917 to 1918. The 6:30pm departure got in 3:30pm the next day, with few rest stops.

Camden shed lasted from 1837 to 1966 with major enlargements taking place by the LNWR in 1920 and the LMSR in 1932.The freight engine roundhouse was closed in 1871, but still exists today as an arts centre.

Left: British Railways designated shed 1A Willesden seen on 25 May 1949. Clustered in the roundhouse are Stanier 5MTs No. 45149 visiting from 8B Warrington, Nos. (4)5025 and 45071, and to the right in light steam Patriot Class No. (4)5502 *Royal Naval Division*. (Online Transport Archive ADav-M032-4)

Below: With Willesden's monolithic concrete coaling tower in the background, Ivatt 2MT 2-6-0 No. 46433 is just five months old on 25 May 1949. This shed was mainly home to the freight workhorses rather than the top link express locomotives. (*Online Transport Archive – ADav-M032-5*)

BR (LMR) Fowler 2-6-2T 3MT No. (400)50 and Ivatt No. 46432 2MT outside the through Willesden shed buildings 25 May 1949. On closure to steam in August 1965 this area subsequently became a Freightliner depot. (*Online Transport Archive ADav-M032-1*)

LMSR 0-6-0 diesel shunter at Camden shed on 12 October 1947. Classed as British Rail D3/7 this was one of the single motored jackshaft drive batch No. 7080–7119. They eventually formed part of Total Operations Processing System (TOPS) Class 11, but the prefix reference was never used. (*Online Transport Archive -Meredith 20-7*)

Unrebuilt Royal Scot Class No. 46140 *The King's Royal Rifle Corps*, Camden, 8 March 1949. The taper boiler, frame and cylinder rebuild was applied in 1952 alongside No. 46158 *The Loyal Regiment* and No. 46164 *The Artists' Rifleman*. (*Online Transport Archive ADav-M017-12*)

Rebuilt Royal Scot No. 46118 *Royal Welch Fusiliers* at Camden on 8 March 1949. Fifteen months after nationalisation, the smokebox and cab-side numbers have changed, but the paired tender of this top link locomotive is still proudly emblazoned with 'LMS'. (*Online Transport Archive – ADav-M017-5*)

Giving the feel of the original locomotives at Camden verses the behemoths of 100 years later. The track and sleeper motif on the brickwork is a nice touch but it is unknown whether this is some artistic licence or a real feature. (*LNWR Postcard, Author's collection*)

Camden sheds and the start of the Camden Incline. This was where steam traction changed to ropework to the station in the earliest years. Later, it was here that carriage brakes were pinned down and tickets collected from the non-corridor stock. Also, it is where passengers would nip out for a call of nature, making it a fragrant area. The winder-style of turnout changer and the turnout switch are interesting to note. (*Contemporary Engraving, 1839*)

The LNWR Primary and Sub Engine Sheds in 1922

(Note: there was no #11 or #36 allocated at this time)

- #1 Camden
- #2 Willesden (sub Watford)
- #3 Bletchley (subs Leighton, Cambridge, Banbury, Oxford, Aylesbury, Bedford, Newport Pagnell)
- #4 Nuneaton (subs Coalville, Loughborough, Leicester, Overseal)
- #5 Northampton
- #6 Bescot (sub Wendsbury)
- #7 Netherfield and Colwick (sub Sheffield)
- #8 Rugby (subs Coventry, Warwick Milverton, Market Harborough, Stamford, Peterborough, Seaton)
- #9 Walsall (sub Dudley)
- #10 Aston (Birmingham) (sub Monument Lane)
- #12 Burton
- #13 Bushbury
- #14 Stafford
- #15 Crewe North Shed (subs Crewe South Shed, Whitchurch)
- #16 Longsight (subs Buxton, Lees, Stockport, Altrincham, Cromford and High Peak)
- #17 Farnley and Wortley
- #18 Birkenhead (subs Hooton, Ellesmere Port)
- #19 Chester
- #20 Huddersfield
- #21 Bangor (subs Caernarvon, Llanberris, Amlwch)
- #22 Holyhead
- #23 Warrington (subs Arpley, Over and Wharton)
- #24 Sutton Oak
- #25 Spring Branch Wigan
- #26 Edge Hill
- #27 Preston (subs Lancaster, Carnforth)
- #28 Tebay (subs Oxenholme, Windermere, Ingleton)
- #29 Carlisle
- #30 Salop (subs Ludlow, Coalport, Craven Arms, Clee Hill, Builth Road, Knighton, Trench)
- #31 Abergavenny (subs Hereford, Tredegar, Blaenavon)
- #32 Workington (sub Penrith)
- #33 Swansea Victoria (subs Camarthen, Llandovery)
- #34 Particroft (sub Plodder Lane)
- #35 Speke Junction (sub Widnes)
- #37 Mold Junction
- #38 Llandudno Junction (subs Corwen, Denbigh, Rhyl)

At the other end of the HS2 first phase, archaeological excavations were taking place at Birmingham Curzon Street, the original terminus station of the Euston line of 1838. One of the surprise elements was the reveal of the original fifteen-locomotive roundhouse. It demonstrates the diminutive nature of Robert Stephenson's early locomotives. The initial roundhouse at Chalk Farm would have been similar. (*HS2 Ltd*)

The original wooden stage (platform) supports, looking like a Roman hypocaust, alongside the terminus rail lines of Curzon Street. This area also includes small carriage and wagon turntables to allow stock to be easily shunted between lines. The early Euston station had the same approach with over forty such turntables to swiftly redisposition stock by horse and hand. (*HS2 Ltd*)

Close-up of the small turntables with central gimbal and holes where the rotational support rail would have been spiked into place. (*HS2 Ltd*)

At the other end of the line to the 1838 Euston station, and designed by Philip Hardwick, was Curzon Street Birmingham, which is close to the roundhouse excavations. It became a goods station and despite being burned out in the Second World War the building has survived and today is Grade I listed. It is being preserved as part of the new HS2 station complex. The columns here are of the Ionic order as opposed to Euston's Doric. (*Tony Hisgett CC BY 2.0*)

The Club put together a scene on the OO Deeping Road MPD layout to represent the mechanical coaling and ash plants. (*Author*)

10
The End and a New Beginning

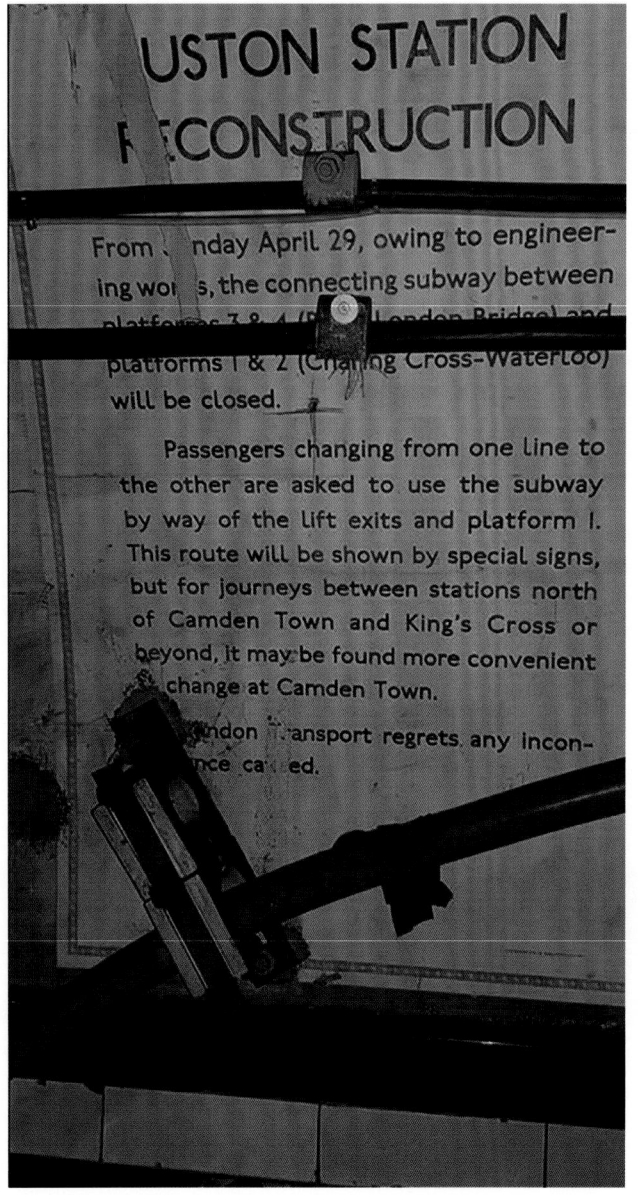

Left: 'Dear commuter, your somewhat gloomy and confusing station is about to be transformed from a complicated Victorian edifice into a brutalist corporate design.' The chaos that ensued in keeping a station open during a full rebuilding, combined with the final stages of west coast electrification, cannot be overstated. The poster seen here is one of several preserved inside Euston Square Underground station, within service tunnels that once had passenger footfall. Image taken during an organised London Transport Museum tour, which can be greatly recommended. (*Author*)

Opposite: While the demolition of the Euston Arch, Great Hall and ancillary buildings was regarded as a pivotal moment in popular regard for Victorian railway infrastructure in 1963, the original station may have disappeared much earlier than thought. This plan shows a 1900 land proposal for the station for a project that was never started. Under the LMSR an American style 'Grand Central' style plan was advanced in 1937 by Percy Thomas, which may have seen both Euston and St Pancras amalgamate. In 1938 substantial facing limestone was extracted for the new building and an accommodation build was begun to rehouse those that would be affected. The Second World War promptly shelved these plans. The white edifice of the Queen's Hotel in Leeds is regarded as the closest to the finished design architecturally. (*LNWRS DBLDG0876A*)

THE END AND A NEW BEGINNING • 95

When the BTC proposed demolition in 1960, Woodrow Wyatt MP tabled a motion to save the Arch and Great Hall. The LCC stated the Arch and lodges should be re-sited. Because the BTC and LCC were in discussion, the Grade II listing was not enforced by a building preservation order. Despite of vigorous protests by many eminent people the Arch was condemned in October 1961 by Ernest Marples, supported by the Conservative Prime Minister Harold Macmillan.

The Architectural Review issued a scathing critique of the Arch demolition in April 1962.

> Its destruction is wanton and unnecessary – connived at by the British Transport Commission, its guardians, and by the London County Council and the Government, who are jointly responsible for safeguarding London's major architectural monuments, of which this is undoubtedly one. In spite of [...] being one of the outstanding architectural creations of the early nineteenth century and the most important – and visually satisfying – monument to the railway age which Britain pioneered, the united efforts of many organisations and individuals failed to save it in the face of official apathy and philistinism.

The breakers moved in on 6 November 1961, the contract for building the new station having been awarded to Taylor Woodrow Construction Limited earlier that year.

As the hotels were demolished and the vista reopened for the first time since the late 1870s, the Arch itself was the first part of the main station to face removal. The three London Underground lifts nearby were time-expired and required replacement as an early step in the rebuilding project. The location of the Arch became the new stair and escalator access point and was deemed the first element of the station to be demolished ready for excavation.

The electrification of the WCML was first proposed in the 1950s. As a project that was accompanied by station rebuilds, track rationalisation and introduction of new rolling stock it represented a large national investment,

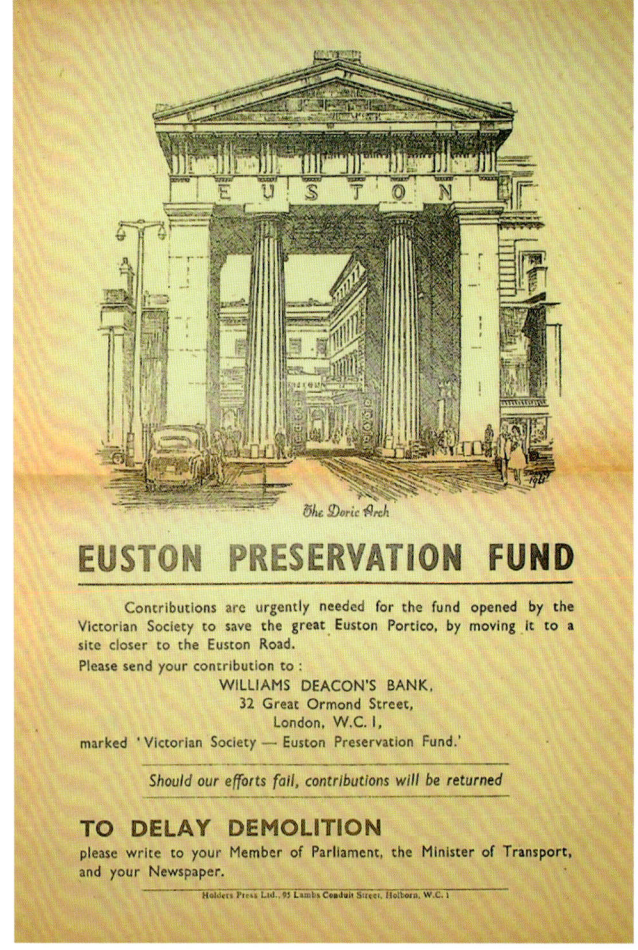

Founded in 1958 to protect Victorian and Edwardian architecture and design, the Victorian Society launched a campaign to save the Euston Arch. In 1962, the Society established the Euston Arch Preservation Trust, which later became the Euston Arch Trust. The trust worked to raise public awareness of the Arch's historical and architectural significance and campaigned for its restoration. Despite failure, these attempts were instrumental in raising public awareness of the importance of preserving Victorian architecture, and it helped to establish the Victorian Society as a leading voice in the preservation of historic buildings and landmarks. (*LNWRS BLDG0009*)

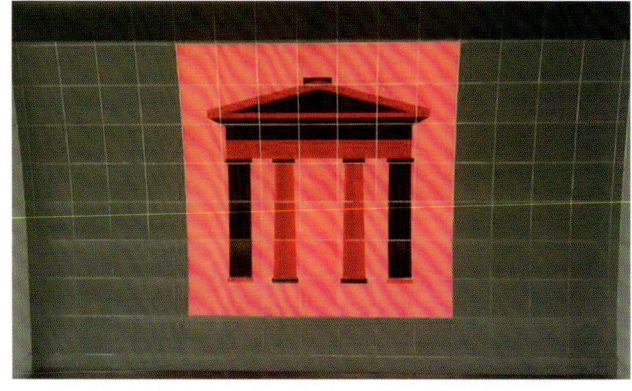

The Victoria Line platforms of December 1968 still commemorate the Arch in pop culture tiling. (*Duane Brown CC BY-SA 2.0*)

Electrification crept south, eventually reaching Euston as reconstruction was in its endgame. For a short time, wires and steam coexisted before a full ban came into play. Here in July 1965 is Stanier 4-6-0 5MT No. 45292 which did not have the diagonal ban stripe on the cab side, allowing it under energised wires south of Crewe. Electric trains started in November of that year. (*Ben Brookbank CC BY-SA 2.0*)

similar to the new HS2 project redeveloping this station in the twenty-first century.

Euston station had become run down, outdated and regarded in strategic planning circles as inadequate to meet the needs of modern rail travel. The rebuild was designed by the architect Richard Seifert and involved the construction of a new, modernist-style building in its place. The new station was characterised by its use of reinforced concrete and steel, with a distinctive 1960s-era design that featured a concourse topped by a large, curved roof. Above the extended platforms was a secondary floor for parcels traffic.

The rebuild of Euston was part of a broader trend in the mid-twentieth century towards the modernisation of infrastructure and the adoption of new technologies and materials. The new station was intended to be more efficient and functional than its predecessor, with improved facilities for passengers and staff.

The end, after 124 years. Harold Macmillan endorsed the decision of Minister of Transport Sir Frank Markham to demolish the Arch, in a written reply to Charles Wheeler, the President of the Royal Academy, in November 1961. This was on the grounds of expense and project delay if funds were to be raised to move it. While the BTC stated £190,000 costs, an alternative scheme quoting less than half, by using rollers was rejected. As can be seen here, scaffolding is being erected by Valori, a Fairclough subcontractor, ready for comprehensive demolition rather than dismantling. Much of the stone fabric ended up in the river Lea's Prescott Channel. Some was used to assist the build of a house in Bromley for company head Frank Valori (notably after he had offered to store the stone for re-erection at his own expense and been rejected), the rest was used for the garden rockeries of associated worthies. Seen here on 28 October 1961, the end is not far off! (*Online Transport Archive Meredith 486-1*)

Above left: The preserved set of main gates at the NRM, part of the Science Museum Collection. (*Timloliver Wiki Commons*)

Above right: The original gates to Euston station from the 1830s are a remarkable survival and now form a display within the town of Shildon. (*Author*)

The reason for the gates' survival was the economy-minded LNWR. As can be seen here one of the Royal Mail access portals from the Doric Arch courtyard to Platform 6 had these gates reused and under cover. Here No. 46256 *Sir William Stanier* drifts to a halt. (*Lens of Sutton Association – PFidczuk B1696*)

From the same vantage point as the previous image. Taken in 1963 with the demolition creeping closer to the remaining buildings. The steel supports for the new Euston are going into place for Platforms 1 and 2. The original cast extensions to make the columns higher are clearly visible. (*Ben Brooksbank CC BY-SA 2.0*)

A feeling of early summer café society is exuded over Euston Grove on 27 May 2022. Soon expansion of temporary HS2 traffic controls will overtake this area. (*Author*)

In the post-war period up to the major station rebuilding, the area between the carriageways of Euston Grove was used as an overflow parking area, along with local bomb sites. Here duplicated on the Club model. (*Author*)

Every effort was made to make Euston feel like a flagship station of the future. As seen here on Platform 10, shiny surfaces look like rain has fallen – in actuality it is highly polished terrazzo flooring. The DC power rails for the Watford services still exist, despite the overhead lines also being present. (*Ben Brookesbank – CC BY-SA 02*)

The Underground still needs ventilation, so a new building nicknamed the 'sugar cube', using more than 13,000 ivory-coloured glazed terracotta faience tiles replaces the Leslie Green structure. These are like those used at Covent Garden and Great Portland Street stations, the aim being to reflect light to the surrounding streets. (*HS2 © WestonWilliamson+Partners_EustonShaft_Cam01*)

Architect's model of the HS2 station new build, north orientated to the left. The removal of the St James's burial ground and the Temperance Hospital all form a part of the evolution of the site, continuing the previous growth activities of the past. (*HS2-VL-39827-220513_HS2_EustonMinisterModel_MM_18 - Middleton Mann © HS2 Ltd*)

The scale of works to the west of the existing station as viewed from the south in April 2022. The bus station occupying the western gardens can be seen, and the red and white barrier line delimits Cardington Street before it is subsumed. (*In The Dark Productions © HS2 Ltd*)

The Euston Plaza in February 2023 and the site of the frontage of the 1963 main station building. This approximates to the position of Euston Street at the rear of the model's Regency villas. The statue of Robert Stephenson by Carlo Marochetti was re-erected in the station plaza. The plans of 1900 and 1937, to swallow all land up to the Euston Square parks, and that executed in 1963, were all aided by the presence of post Blitz bomb sites, since far less compulsory purchase overhead was required. The jet-age station was designed internally by unnamed British Railway architects, and the frontage of enclosing offices by Richard Seifert, he of Centre Point and NatWest Tower fame. (*Author*)

The main concourse of Euston station, the modern equivalent of the Great Hall and much better placed across the platform entrances rather than between platforms. Drummond Street and the Euston Arch approximate to this façade. This ably demonstrates the difference between the organic piecemeal development of the original station and the ability to wipe the slate clean. While some calls were made for HS2 to include wholesale replacement of this iteration of the station, it is not yet life expired. At some time in the future there will be calls to preserve this as an example of its kind. (*Author*)

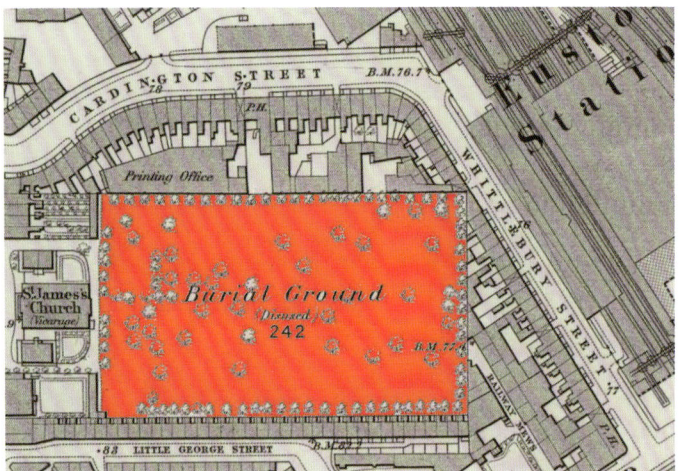

As can be seen here on the 25in to the mile 1876 map, the growth of Euston station expanded east, to create Platforms 1 and 2, and west with a new, almost independent, station alongside Whittlebury Street. To the north of Cardington Street were carriage sidings manually shunted over numerous turntable platforms. As the physical size of a carriage wheelbase grew these became redundant, and a new facility was opened further to the north.

By the time of the 1915 map, the arrivals station and Cardington Street deviation had cropped the corner off the St James's burial ground, as the original valley sides were cut away and protected by new brick revetments. To the north a locomotive turntable was now in place. Whittlebury Street alas is no more. Little George Street has moved up a generation from Regency, to a Victorian Coburg street appellation.

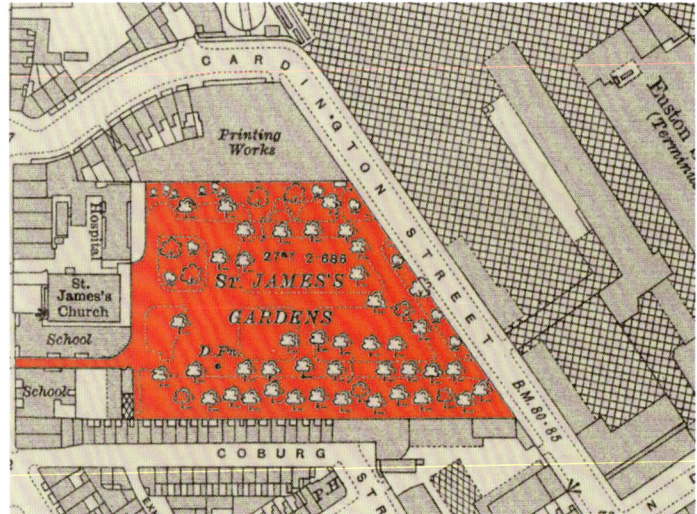

The St James's burial ground was full and closed to new burials at the time of the Euston station foundation. It was subsequently cleared of headstones and memorials to make an area of parkland. To show how St James's would have looked originally are these two images of Bunhill Fields burial ground. Situated near Bishopsgate and adjoining the Honourable Artillery Company it is well worth a visit. Surviving memorials such as that of author Daniel Defoe echo how one for Captain Matthew Flinders would have stood out in a crowded St James's. (*Author*)

THE END AND A NEW BEGINNING • 103

While the St James's burial ground had closed nearly twenty years prior to the model, we could not resist putting in a funeral cortège along Euston Square west. These are white metal OO hearse and landaulette/landau carriage kits from Langley.

The MOLAS (Museum of London Archaeology Service) excavation of the St James's site in preparation for the HS2 station works. This plot of land adjacent to Euston was bought in 1788 by St James's Church Piccadilly for use as an extra burial ground for the parish. It was in use between 1790 and 1853. Around 57,000 bodies were thought to have been buried here. Careful exhumation was followed by reburial at Brookwood in Surrey. Notably this was also the destination of the London Necropolis Railway. (*HS2 04/06/19*)

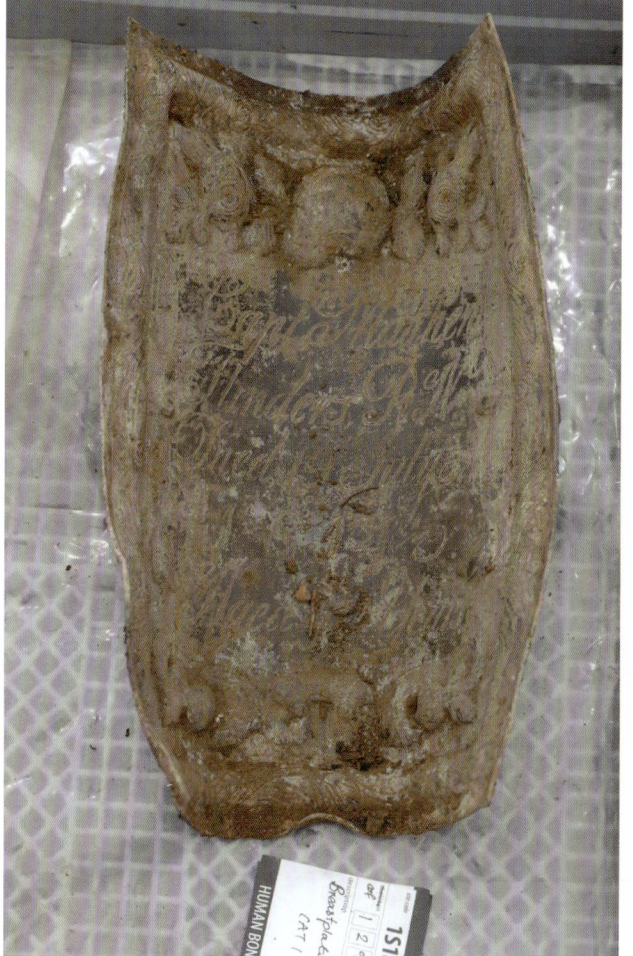

Left: Several notable burials were known to have been made in St James's, including naval explorer of Australia, Captain Matthew Flinders. A concern was that chopping off the north-west corner in the Victorian expansion resulted in the loss of this grave plot. (*HS2 Ltd James O. Jenkins*)

Below: The exposure of the grave was regarded as a major milestone in the project. (*HS2 Ltd James O. Jenkins*)

104 • EUSTON: A HISTORY AND MODELLING THE 1875 STATION

Removal of the foundation stone belonging to the London Temperance Hospital at the entrance to the St James's burial ground. When it was laid on Thursday, 24 April 1884 by the Duke of Westminster, a time capsule was secreted inside. (*HS2 Ltd*)

Just 133 years later a Fikheel anti-neuralgia toothpaste advert is the first thing to be seen (oil of cloves and cocaine). The contents included the following: the *Temperance Chronicle*, 12 April 1884; the *Alliance News*, 12 April 1884; the *Medical Temperance Journal*, April 1884; and the *Tenth Annual Report of the London Temperance Hospital*, 29 May 1883. (*HS2 Ltd*)

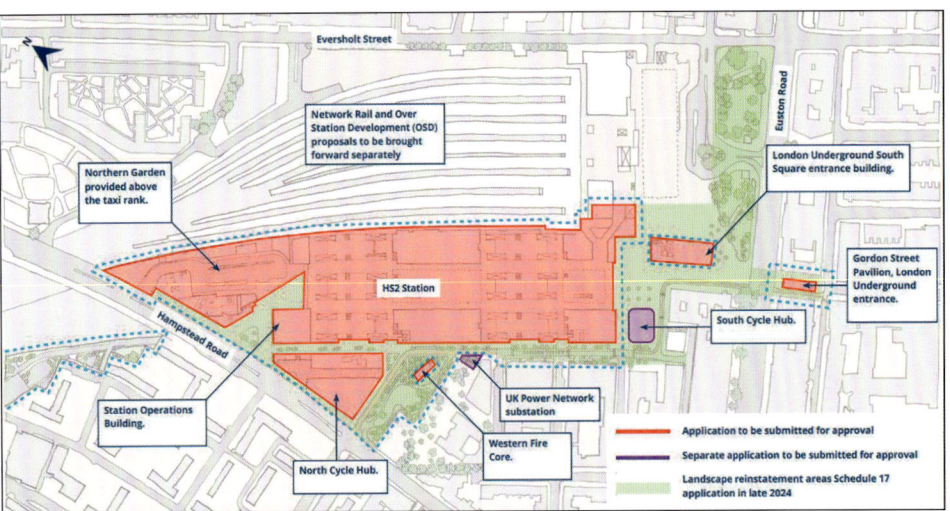

Euston, we have a problem! The most recent overall plan for the site before government stops were put on the current Euston redevelopment project on cost grounds early in 2023. Already reduced from eleven to ten platforms, in what manner the flagship will be completed remains to be seen. Will Old Oak Common be the new superhub and Euston a branch line? (*HS2 Ltd*)

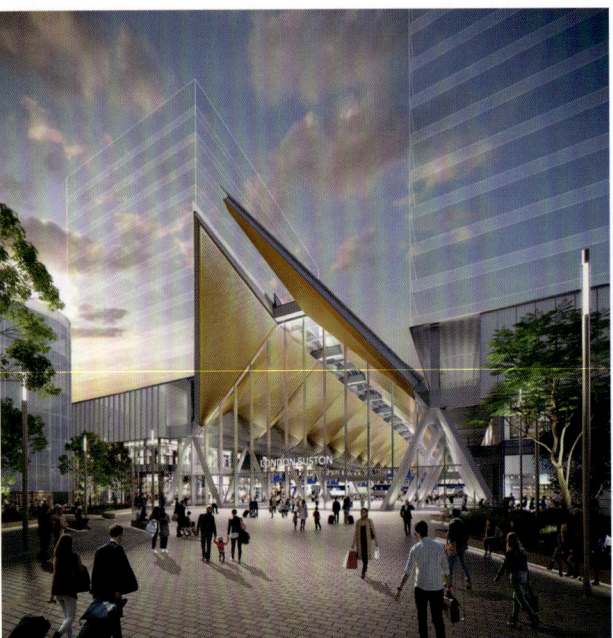

Left and below: Different from either iteration of Euston station so far, the HS2 station as projected. As with many original Tube stations, the future ability to build offices above the buildings is recognised. (*Grimshaw HS2 –VL-36170/1-Feb22*)

11
The Sultan of Zanzibar's Visit

Treaty between Her Majesty and the Sultan of Zanzibar for the suppression of the Slave Trade.

In the name of the Most High God

HER Majesty the Queen of the United Kingdom of Great Britain and Ireland, and His Highness the Seyed Burgash-bin-Saîd, Sultan of Zanzibar, being desirous to give more complete effect to the engagements entered into by the Sultan and his predecessors for the perpetual abolition of the Slave Trade, they have appointed as their Representatives to conclude a new Treaty for this purpose, which shall be binding upon themselves, their heirs, and successors, that is to say, Her Majesty the Queen of Great Britain and Ireland has appointed to that end John Kirk, the Agent of the English Government at Zanzibar, and His Highness the Seyed Burgash, the Sultan of Zanzibar, has appointed. to that end Nâsir-bin-Saîd, and the two aforesaid, after having communicated to each other their respective full powers, have agreed upon and concluded the following Articles ... (*NMM collection*)

Dickens could have portrayed Somers Town in the hours prior to the Sultan's arrival in this way:

> The appearance presented by the streets of London an hour before sunrise, on a summer's morning, is most striking even to the few whose unfortunate pursuits of pleasure, or scarcely less unfortunate pursuits of business, cause them to be well acquainted with the scene. There is an air of cold, solitary desolation about the noiseless streets which we are accustomed to see thronged at other times by a busy, eager crowd, and over the quiet, closely-shut buildings, which throughout the day are swarming with life and bustle, that is very impressive. The last drunken man, who shall find his way home before sunlight, has just staggered heavily along, roaring out the burden of the drinking song of the previous night; the last houseless vagrant whom penury and police have left in the streets, has coiled up his chilly limbs in some paved corner, to dream of food and warmth. The drunken, the dissipated, and the wretched have disappeared; the more sober and orderly part of the population have not yet awakened to the labours of the day, and the stillness of death is over the streets; its very hue seems to be imparted to them, cold and lifeless as they look in the grey, sombre light of daybreak. The coach-stands in the larger thoroughfares are deserted: the night-houses are closed; and the chosen promenades of profligate misery are empty.

Sketches by Boz: Illustrative of Every-day Life and Every-day People, Charles Dickens, 1836

The Chosen Event

The Euston diorama in its reskinned and expanded state is aimed at a precise date and time to focus the experience. The summer 1875 visit to Britain by the Sultan of Zanzibar stood out. Since this was a combination of a direct meeting with the British government for trade talks, and the ratification of an anti-slavery treaty, it made this event an ideal candidate, since a ceremonial visit allowed extra details to be presented. Researching the background of this has brought to light a whole miasma of political machinations. The press had a great time with the political and royal angle (some things really never change). It was also a demonstration of the organisational ability of the early railway.

The Station in 1875 Fitted the Model!

The date was prior to the physical joining of the Euston and Victoria hotels by the extension over the roadway – in the creation of Euston Place, the long vista from Euston Square Gardens was blocked. The station itself was in a state

Satirical magazines such as *Punch* had great fun with the sultan's visit, and the prime minister's discomfiture. Both the press and notation in *Hansard* record that the six-week stay had a certain publicly underwritten extravagance that was perhaps disproportionate to the end result.

of rebuilding and expansion, much the same as in the twenty-first century with HS2. There was some canopy rebuilding around this time, so train workings would not be hidden from a viewing public at exhibitions. The station frontage itself was still in its original 1838 form, and the stonework was now weathering darker with pollution. Extended office accommodation had not yet been squeezed into Drummond Street (this was a 1910 to 1920 extension including loss of a logia). Changes of use within buildings had happened due to new technology – for example, one of the logia now housed the telegraph office and telegrams could be sent along the railway for a premium, but the external walls remained the same.

The timing on this bright summer's day in the model is the end of the dress rehearsal for the sultan's arrival. The red carpet is now out and brushed, the dignitaries are gathering to send the august personage on his way by to Birmingham. The sultan visited a number of industrial sites from London and in theory made this into a trade mission to disguise the true forced political intent.

A company of the Coldstream Guards who were quartered in a London barracks is attending officially and by platoon is marching in and up Platform 6. Their band is forming up to play the national anthems of both countries but at this point would be discordantly tuning up amidst the sounds of a working steam mainline station. Access from the main station courtyard has been sealed off and the Royal Mail has to seek a different access point. The station has effectively lost a platform for much of the day until this little piece of pomp has been completed. The sultan will be directed through the doors of the Great Hall corridor and pick up his carriage at the main entrance. Passengers wishing to purchase tickets will be directed around the courtyard to the west and directly into the Great Hall to access the two main ticket offices.

It is interesting to watch old Pathé newsreels of the royal departures of King George VI and Queen Elizabeth II for excursions by ship for a state visit or arrival back from a Scottish holiday at Euston. Royal activity was normally this side of the station with its wider platforms. For all the seeming slickness and speed involved it also shows the organised chaos of isolating a section of station: in the background was the changing the signalled pathing for other trains, making sure everything was scrubbed clean or hidden, controlling the public and maintaining a police protective presence. The standard station public control was a succession of low

The original diorama had no platforms, it was all front of house buildings, so our expansion began on the easternmost baseboard exposure. Once the theme was chosen, we printed contemporary anti-slavery posters to place on the arrival platforms of the sultan's train. No factual evidence exists for this, but the Wilberforce campaign would have made the most of every opportunity. (*Author*)

wooden barriers on wheels that were used to divide platforms when multiple trains were formed up. These existed up until the 1960s and were used on royal duty.

Transcribing this back to mid-Victorian times and a royally vetoed visit, government pool open landaus would have been sourced for the dignitaries and lesser worthies in place of the carriages of the Royal Mews. There would then be closed two-horse carriages for associated major domos and the household of the sultan. Beyond that there would also have been a substantial shipment of baggage. The probability would be that the railways would use their normal parcels wagons to deliver promptly from the train to the hotel in Mayfair that was to be the main residence.

Meanwhile, in a form of organized chaos, the railway's own police force at Euston would have been supplemented for the day from elsewhere on the system, plus London Bobbies seconded for crowd control. Passengers with disrupted services would be informed by chalkboards and naval-style megaphones (no wired tannoy systems), perhaps accompanied by a grudging thumb over shoulder and an 'over there mate' from a fed-up porter. Tempers would be short due to clearing the platform of mailbags and crates, only to have to get it all out again as soon as the personage and 'his nibs' the station master had gone.

Interestingly, film newsreels show a later Sultan of Zanzibar at Euston station en route to exile in Southsea, having been overthrown in 1963, to be followed in the next year by the refugee parents of Freddie Mercury of Queen fame.

When choosing the visit of the sultan to ratify the treaty as the core display on the Euston model, little did we know of the controversial background splitting Crown and State. The newspapers surrounding the event were researched online at the British Newspaper Archive, which is located at https://www.britishnewspaperarchive.co.uk/. This material was used alongside the political recordings of the House of Commons proceedings from *Hansard* and satirical magazines of the day.

Gunboat diplomacy was fully utilised, with HMS *London* stationed off the African coast and centred on the island of Zanzibar and its dependencies. The desire to cease international slavery culminated in an 1807 Act of Parliament. However, the law had become blurred by the

Seen in top hat and sash and present to represent the government, but not the queen in this event, is His Grace Arthur Wellesley, 2nd Duke of Wellington, shown in a satirical print from *Vanity Fair* of 1872. At this time, he was the Lord Lieutenant of the County of Middlesex. He was accompanied by his wife, Lady Elizabeth. The duke was to die in August 1884 within the confines of Brighton railway station.

The ceremonial band gathers as a final rehearsal takes place to ensure that the red carpet is correctly located. (*Author*)

Palmerston administration support of the US Confederate cotton slave states in the 1860s, attempting to support the cotton industry. Now, essentially as a spoiler for the German and French colonial rush for Africa, Britain could implement strict policy. The aim was to prevent cheap labour being applied to, and slaves being exported from, a number of East African countries.

Dr David Livingstone had earlier published accounts of the slave trade in that region, and this had caught the popular imagination in British newspapers. With the aid of John Kirk, the government political agent in the region, Prime Minister Benjamin Disraeli was able to effectively use carrot (trade deals) and stick (military action) to impose a treaty on Barghash bin Said, the Sultan of Zanzibar. This forced the closure of all slave markets, ceased the traffic of slaves, and forbade British and Indian citizens in the country from owning slaves.

The sultan arrived in Britain by steamer at Gravesend (annoying for the Euston modellers, since a boat train from Liverpool would have been a larger occasion). In real life the desire was to encourage trade, and a visit to the Midlands on Friday, 2 July provided an opportunity to further this. The 10:10am departure to Birmingham would entail a reserved train, and a smaller scale send-off to get the Sultan in a receptive mood. When he arrived in Birmingham at 1pm the city aldermen were there to greet him.

This treaty and trade visit was purely a governmental affair – the royal family had effectively vetoed it. This said, the Prince of Wales still held a small reception at Marlborough House on Monday, 14 June. Royal rejection is said to be due to the maltreatment of a distant female cousin, who had changed to the Christian faith. So, while there was some pomp provided, it was more of a dry governmental style. A series of society parties were attended, and a fireworks display held at the Crystal Palace by Messrs Brock and Co. on Saturday, 19 June. The queen would not give way, so the sultan's trip to Royal Ascot resulted in a press frenzy, due to the royal enclosure being closed to him.

Above: The baseboard with the military display is the only one that is not transported as a flat item. Therefore, an enveloping wooden shield has been created. This will also guard the canopy cutaway which will be the final stage of the station build (after this book). The kiosk is a nod to the first ever WH Smith bookstall opened on 1 November 1848. (*Author*)

Right: The LNWR warranted police team begin to clear part of the Great Hall to allow the sultan and his retinue to progress from carriages in the main courtyard, through the furthest right-hand door and onto the red carpet on the other side of the eastern office block. The module has been orientated to get the mid-morning sun from the south. (*Author*)

APPENDICES

Appendix A

The Unratified Anti-slavery Treaty

Treaty between Her Majesty and the Sultan of Zanzibar for the Suppression of the Slave Trade.

In the name of the Most High God
HER Majesty the Queen of the United Kingdom of Great Britain and Ireland, and His Highness the Seyed Burgash-bin-Saîd, Sultan of Zanzibar, being desirous to give more complete effect to the engagements entered into by the Sultan and his predecessors for the perpetual abolition of the Slave Trade, they have appointed as their Representatives to conclude a new Treaty for this purpose, which shall be binding upon themselves, their heirs, and successors, that is to say, Her Majesty the Queen of Great Britain and Ireland has appointed to that end John Kirk, the Agent of the English Government at Zanzibar, and His Highness the Seyed Burgash, the Sultan of Zanzibar, has appointed. to that end Nâsir-bin-Saîd, and. the two aforesaid, after having communicated to each other their respective full powers, have agreed upon and concluded the following Articles: -

ARTICLE I. **THE** provisions of the existing Treaties having proved ineffectual for preventing the export of slaves from the territories of the Sultan of Zanzibar in Africa, Her Majesty the Queen and His Highness the Sultan above named agree that from this date the export of slaves from the coast of the mainland of Africa, whether destined for transport from one part of the Sultan's dominions to another or for conveyance to foreign parts, shall entirely cease. And His Highness the Sultan binds himself, to the best of his ability, to make an effectual arrangement throughout his dominions to prevent and, abolish the same. And any vessel engaged in the transport or conveyance of slaves, after this date, shall be liable to seizure and condemnation by all such naval or other officers or agents, and such Courts, as may be authorized for that purpose on the part of Her Majesty.

ARTICLE II. **HIS** Highness the Sultan engages that all public markets in his dominions for the buying and selling of imported slaves shall be entirely closed.

ARTICLE III. **HIS** Highness the Sultan above named engages to protect, to the utmost of his ability, all liberated slaves, and to punish severely any attempt to molest them or to reduce them again to slavery.

ARTICLE IV. **HER** Britannic Majesty engages that natives of Indian States under British protection shall be prohibited from possessing slaves and from acquiring any fresh slaves in the mean time [the words 'in the meantime' are redundant here. They were connected in the original English draft and in my translation, from which they are copied, with the sentence 'from and after a date to be hereafter fixed' – G.P.B.] from this date.

ARTICLE V. **THE** present Treaty shall be ratified and the ratifications shall be exchanged at Zanzibar as soon as possible, but in any case in the course of the 9th of Rabîa-el-Akhir [5 June 1878] of the months of the date hereof. In witness whereof the respective Plenipotentiaries have signed the same, and have affixed their seals to this Treaty, made the 5th of June, 1878, corresponding to the 9th of the month Rabîa-el-Akhir, 1290.

(Signed) **JOHN KIRK**,
Political Agent, Zanzibar. (L.S.)

The mean in God's sight,

(Signed) NASIR-BIN-SAID-BIN-ABDALLAH

With his own hand.

[No seal is appended to this signature. The defect is made good by the signature and seal of the Sultan to the ratifications following. – G.P.B.]

We have looked into and considered this Treaty, and we agree to it and accept it; and we confirm everything which it sets forth in all its provisions and articles. And we confirm the same on behalf of our heirs and those who may succeed us, giving our firm bond and covenant, and our faithful word, to carry out all that is set forth in the body of this written document, and to avoid as much as possible everything that contravenes it, and to the best of our ability not to transgress its provisions and conditions. In conformation of which we hereto affix our seal and our signature with our own hand this 9th of Rabîa-el-Akhir, 1290 [5 June 1875].

Approved by
The poor, the unworthy,
(Signed) **BARGHASH-BIN-SAID-BIN-SULTAN**.

Written by his own hand. (L.S.)
Translated by

(Signed) **George Percy Badger**.
June 30, 1875.

Appendix B
Slavery Treaty Ratification

Treaty between Her Majesty and the Sultan of Zanzibar, supplementary to the Treaty for the Suppression of the Slave Trade of June 5, 1875.

Signed at London, July 14, l875.

HER Majesty the Queen of the United Kingdom of Great Britain and Ireland, and His Highness the Seyyid Barghash-bin-Said, Sultan of Zanzibar, having concluded a Treaty at Zanzibar on the 5th June, 1875, corresponding to the 9th of the month of Rabîa-el-Akhir, A.H. 1290, for the abolition of the Slave Trade, and whereas doubts have arisen or may arise in regard to the interpretation of that Treaty, Her Britannic Majesty and His Highness the Sultan of Zanzibar have resolved to conclude a further Treaty on this subject, and have for this purpose named as their Plenipotentiaries, that is to say:

HER Majesty the Queen of the United Kingdom of Great Britain and Ireland, the Right Honourable Edward Henry, Earl of Derby, Baron Stanley of Bickerstaffe, a Peer and a Baronet of England, Her Majesty's Principal Secretary of State for Foreign Affairs, &c., &c., &c.;

And His Highness the Seyyid Barghash-bin-Said, Sultan of Zanzibar, Nâsir-bin-Saîd-bin-Abdalla;

Who, after having communicated to each other their respective full powers, have agreed upon and concluded the following Articles:

ARTICLE I. **THE** presence on board of a vessel of domestic slaves in attendance on or in discharge of the legitimate business of their masters, or of slaves bona fide employed in the navigation of the vessel, shall in no case of itself justify the seizure and condemnation of the vessel, provided that such slaves are not detained on board against their will. If any such slaves are detained on board against their will they shall be freed, but the vessel shall, nevertheless, not on that account alone be condemned.

ARTICLE II. **ALL** vessels found conveying slaves (other than domestic slaves in attendance on or in the discharge of the legitimate business of their masters, or slaves bona fide employed in the navigation of the vessels) to or from any part of His Highness' dominions, or of any foreign country, whether such slaves be destined for sale or not, shall be deemed guilty of carrying on the Slave Trade, and may be seized by any of Her Majesty's ships of war and condemned by any British Court exercising Admiralty jurisdiction.

ARTICLE III. **THE** present Treaty shall be ratified, and the ratifications shall be exchanged at Zanzibar as soon as possible.

In witness whereof the respective Plenipotentiaries have signed the same, and have affixed thereto their seals.

Done at London the fourteenth day of July, in the year of Grace one thousand eight hundred and seventy-five.

(LS) **DERBY.**
(LS.) **NASIR-BIN-SAID-ABDALLAH.**
This is ratified.
(LS.) **BARGHASH-BIN-SAID.**

* The Sultan of Zanzibar's Ratification is attached to the original Treaty. That of Her Majesty was delivered to the Sultan in Zanzibar, September 20, 1875.

Appendix C

Hansard Parliamentary Extracts for the Sultan's visit

Hansard, Vol. 221, 4 August 1874

SIR WILFRID LAWSON
asked the Under Secretary of State for Foreign Affairs, Whether the Sultan of Zanzibar, since the signing of the Treaty in May last year, has not, through Her Majesty's Representative, intimated a wish to visit England; and, if so, whether Her Majesty's Government will not, in view of furthering the object of that Treaty, deem it expedient to facilitate such a visit?

MR. BOURKE
I understand that the Sultan of Zanzibar has upon one or more previous occasions intimated his wish to pay a visit to England; but no communication to that effect has reached Her Majesty's present Government. Two or three days ago a letter was received from the Sultan of Zanzibar; but there has not yet been sufficient time to ascertain Her Majesty's wishes upon the subject.

Hansard, Vol. 225, Thursday, 15 July 1875

SIR JOHN KENNAWAY
asked the Under Secretary of State for Foreign Affairs, Whether any communications have lately passed between Her Majesty's Government and His Highness the Sultan of Zanzibar, with respect to the suppression of the Slave Trade within his dominions; and, if so, whether he has any objection to communicate the substance of them to the House?

MR. BOURKE
, in reply, said, that communications had passed between Her Majesty's Government and the Seyyid of Zanzibar since he had been in this country. The House was aware that, under the Treaty of 1873, certain doubts had arisen which the Law Officers of both Her Majesty's present and late Government considered would impair the efficiency of the Treaty. Another Treaty had been signed by the Sultan since he had been in this country; and he (Mr. Bourke) trusted that the result of that would be that the object of the Treaty of 1873 would now be carried out in its full force. There would be no objection to lay the Treaty before the House in a few days as soon as necessary formalities had been gone through.

Hansard, Vol. 226, Wednesday 4 August 1875

(36.) £ 7,500, for the Entertainment of the Sultan of Zanzibar.

(37.) £ 8,250, for Repayment of Moneys under "The London and North Western Railway Company (New Lines, &c.) Act, 1875."

Her Majesty's speech for prorogation of parliament - delivered by the Lord Chancellor on Friday 13 August 1875.

"My Lords, and Gentlemen,

I am happy to be enabled to release you from your attendance in Parliament.

The relations between myself and all Foreign Powers continue to be cordial, and I look forward with hope and confidence to the uninterrupted maintenance of European peace.

The visit paid to this country, on the invitation of my Government, by the Ruler of Zanzibar, has led to the conclusion of a Supplementary Convention, which, I trust, may be efficacious for the more complete suppression of East African Slave Trade."

Appendix D

Hansard Parliamentary Extracts for the Euston Area

Hansard, Vol. 169, debated Thursday, 12 March 1863

The Metropolitan, Hampstead, and Tottenham scheme—which proposed to occupy the whole north side of Euston Square with a station, and against the inconvenience to which they would thus be subjected he had been requested on the part of the inhabitants to protest. The railway to which he referred was not, like the Great Eastern, promoted by a large company, but was a speculation rather of a certain number of engineers and contractors. Having a station in the square, it was proposed that it should pass under a considerable extent of ground occupied by the houses of the poor—thus inflicting upon the inhabitants of the district all those evils which had been so graphically depicted by the noble Earl who had just spoken. Under those circumstances, he thought it would be well if the Government would consent to the postponement of the Bill for the construction of that railway, as well as of the other Bills by means of which it was proposed to effect similar objects, for a year. Now, that the metropolis seemed literally about to be taken by storm, the subject was one which demanded their serious attention. If it were necessary that we should have these railways through London, they should take care that it should be done with a minimum of evil.

Drainage of Towns, *Hansard*, Volume 56, debated Friday, 12 February 1841

There was an instance in illustration of the facility with which the evil was abated when the cause was removed, given before the committee of the Commons last year, and no doubt familiar to their Lordships from the fact that it had been repeatedly given in the newspapers. There was a ditch near the terminus of the Birmingham Railway at Euston-square, into which a number of drains discharged themselves. The consequence was, typhus fever continually prevailed there. The simple remedy was applied of covering in the ditch, and ever since the place had been quite free from such infectious diseases.

Appendix E

Hansard Parliamentary Extracts for the Building of the London and Birmingham Railway

Hansard, Volume 14, debated 9 August 1832

Lord George Bentinck said, he had lately sat on a Committee upon the plan for making a rail-road from London to Birmingham, and it was proved that the loss to the revenue in the abolition of stage-coaches would be 78,000l. annually. The noble Lord could not consider it a violation of principle to tax steam where it injured the interest of stage-coach property.

Colonel Sibthorp could not be forgetful of the interests of his agricultural constituents, suffering severely under the pressure of poor-rates. The use of vehicles impelled by steam would injure those constituents. It would throw greater obstacles than at present in the way of the landed interest.

Mr. Ewart thought that the revenue must have been benefited by the Manchester rail-way. Trade and manufactures had been promoted, and barren land brought into cultivation to a wonderful extent by the conveyance of manure.

PRIMROSE HILL TUNNEL.

Beyond Kentish Town, and under Chalk Farm Bridge, a spot of celebrated duels, the Primrose Hill Tunnel is reached. 'Without a moment's warning, plunged into worse than Cimmerian darkness, and hurried along through clouds of smoke and vapour; amid flying sparks, jarring atoms, rushing winds, and every sign of elemental strife, whilst stunning sounds, and a rattling, clashing din, form a hubbub than which what Satan heard in his flight through the realms of Chaos and Old Night could scarcely be more terrific.' (*Drake's Road Book*, 1839)

Appendix F
1946 Ordnance Survey Photo Mosaic Element for Euston East

Taking in much of Somers Town with Euston Station to the left. Showing some canopy patch repairs from bomb damage. The westernmost part of the Euston Hotel sustained bomb damage and the crescent backing onto it was destroyed. This extended to the mews of the Euston Square villas, some of which sustained blast and fire damage resulting in later demolition in their terraces. Note: This image is colour-adjusted and edge-enhanced from the original. (*CCBY, reproduced with the permission of the National Library of Scotland*)

Appendix G
1946 Ordnance Survey Photo Mosaic Element for Euston West

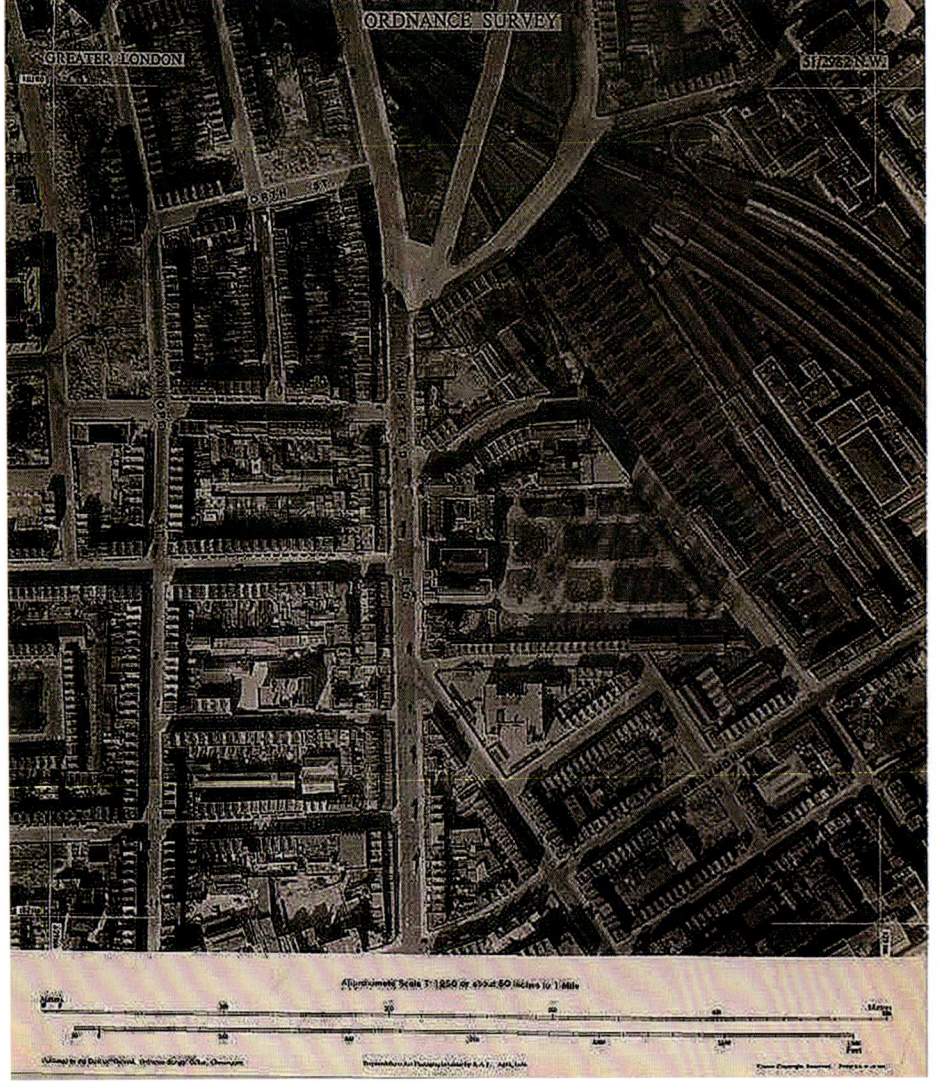

While there were several bombs dropped in this station area, the station itself came out of the war surprisingly intact due to efficient rebuilding. Further to the west, off towards Stanhope Road, both aerial mines and V1 flying bombs caused major damage. Note: This image is colour-adjusted and edge-enhanced from the original. (*CCBY, reproduced with the permission of the National Library of Scotland*)

Bibliography and Useful Sources

Charles Booth Police Notebooks published by the LSE – https://booth.lse.ac.uk/notebooks/police-notebooks

Charles Booth Poor Maps published by the LSE – https://booth.lse.ac.uk/

Drake's Road Book – Grand Junction Railway 1838: Moorland Reprints : Pub 1974

Drake's Road Book – London and Birmingham Railway 1839: Moorland Reprints : Pub 1974

Euston Station Through Time : John Christopher : Amberley Publishing : Pub 2012

Lionel Pincus & Princess Firyal Map Division, The New York Public Library

LMS 150 – Patrick Whitehouse & David St John Thomas : David & Charles : Pub 2002

LMS Miscellany, Volumes 1/2/3 H N Twells : OPC : Pub 1982

London Main Line War Damage : B.W.L.Brooksbank : Capital Transport : Pub 2007

London's Lost Tube Schemes : Anthony Badsley-Ellis : Capital Transport : Pub 2005

London's Secret Tubes – Andrew Emmerson and Tony Beard : Capital Transport : Pub 2004

Lost Victorian Britain : Gavin Stamp : Aurum Press : Pub 2010

Public Record Office census detail 1851–1881 online.

Rails Through the Clay : Desmond F. Croome and Alan A. Jackson : George Allen & Unwin : Pub 1964

Railway Design Since 1830 : Brian Haresnape: Ian Allan Publishing : Pub 1968

Railway Picture Postcards : Maurice I. Bray : Mooland Publishing : Pub 1986

Railway Wonders of the World Magazine: Clarence Winchester/Cecil J. Allen : RCH : Pub 1936

Sixties Spotting Days Around London and the Home Counties: Kevin Derrick : Amberley Publishing : Pub 2016

Stokers and Pokers: LNWR Electric Telegraph and Railway Clearing House : John Murray : Pub 1849

The Blitz Then and Now - Volume 2 : After the Battle : Pub 1990

The Diesel Shunter: Colin J Marsden : OPC : Pub 2003

The London and North Western Railway: Neil Smith : Pen and Sword : Pub 2021

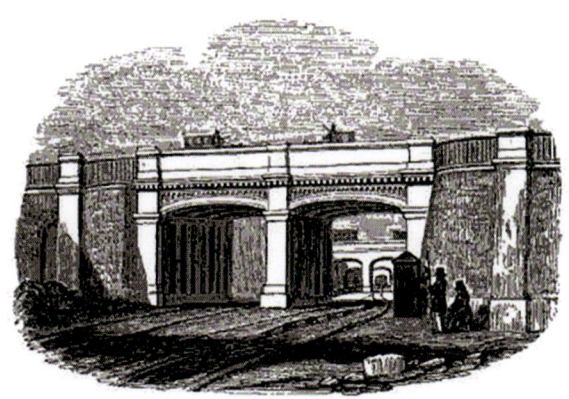

The Park Street Bridge, one of seven of iron and brick construction in the mile long 'Grand Excavation' of the Camden Bank through the London clay at an average of 1 in 85 gradient. (*Drake's Road Book, 1839*)

Much work has gone into the research and model making on this project. In modern life, railway modelling is a hobby that can aid socialisation and prevent things like rural or indeed urban isolation. If you are able, make your way to the local model shows, have a chat, exchange ideas, visit other domestic layouts or join a club. The author has an accompanying LMSR warning horn to blow for attention when required! (*Author's collection*)

Why We 'Do' Railway Modelling

As a registered charity we actively encourage the promulgation of this hobby for a number of reasons.

For the younger railway modeller it stimulates cognitive engagement, fosters critical thinking, enhances problem-solving abilities and promotes creativity. It helps develop a range of cognitive skills that can have long-lasting benefits in various aspects of life, from academic pursuits to professional endeavours and personal growth.

For an older person the mental benefits of railway modelling can be summarised as follows.

- **Cognitive stimulation**: engaging in model-building exercises stimulates the brain and promotes cognitive activity.
- **Memory enhancement**: building models requires learning and recalling information, improving memory retention and retrieval.
- **Focus and concentration**: model building exercises require sustained focus and concentration, helping to improve attention skills.
- **Fine motor skills and dexterity**: manipulating small components in model building activities can help maintain or improve fine motor skills and coordination.
- **Problem-solving and critical thinking**: building models involves solving puzzles and overcoming challenges, stimulating problem-solving and critical thinking abilities.

(Author)

(Author)

(Author)

(Author)